FIREFLY

STARGAZING WITH A TELESCOPE

THIRD EDITION

ROBIN SCAGELL

FIREFLY BOOKS

A FIREFLY BOOK

Published by Firefly Books Ltd. 2009

First printing

Publisher Cataloging-in-Publication Data (U.S.)

Scagell, Robin.
Stargazing with a telescope / Robin Scagell.
3rd ed.
[192] p.: col. ill., photos.; cm.
Includes index.
Summary: Describes the wide range of telescopes that are available internationally, with examples of objects to observe from both northern and southern hemispheres. Also offers advice about accessories such as eyepieces and filters, plus suggestions for astro-photography using cameras, CCDs and webcams.
ISBN-13: 978-1-55407-577-5 (pbk.)
ISBN-10: 155407-577-7 (pbk.)
1. Stars -- Observers' manuals. 2. Telescopes -- Amateurs' manuals. 3. Astronomy -- Amateurs' manuals. I. Title.
522.2 dc22 QB63.S365 2009

Library and Archives Canada Cataloguing in Publication

Scagell, Robin
Stargazing with a telescope / Robin Scagell. -- 3rd ed.
ISBN-13: 978-1-55407-577-5
ISBN-10: 155407-577-7
1. Telescopes--Amateurs' manuals. 2. Astronomy--Amateurs' manuals. 3. Stars--Observers' manuals. I. Title.
QB63.S365 2009 522'.2 C2009-901694-X

Published in the United States by
Firefly Books (U.S.) Inc.
P. O. Box 1338, Ellicott Station
Buffalo, New York 14205

Published in Canada by
Firefly Books Ltd.
66 Leek Crescent
Richmond Hill, Ontario L4B 1H1

Published in Great Britain in 2009 by Philip's,
a division of Octopus Publishing Group Ltd,
2–4 Heron Quays, London E14 4JP

Printed in China

Front cover:
Sky-Watcher SkyHawk 1145 PM (Optical Vision Ltd); First Quarter Moon (Robin Scagell).

Back cover:
Meade LXD 75 (Meade); Horsehead Nebula (Michael Stecker).

Title page:
Stargazer (Robin Scagell).

CONTENTS

INTRODUCTION

This book is for anyone who always wanted an astronomical telescope but was never sure how to go about it, or maybe who has one already and doesn't know how to put it to best use. It is written for the beginner to astronomy, and for that reason I have tried not to make assumptions about what you might know already. With the help of this book you should be able to make sense of the ads you find in the astronomy magazines, and know what all the bits and pieces are for. I hope I can guide you through some of the pitfalls that may result in making a wrong choice. There are some things I can't cover, particularly regarding the quality of instruments from particular manufacturers, but I hope I have given enough detail for you to make up your own mind.

When it comes down to it, telescopes are very individual things. The optics are usually hand-finished, and they must be aligned accurately within the tube – which leaves plenty of room for things to go wrong. If you get a bad telescope, you should expect it to be replaced – but you will have to know what you are talking about. Even the best telescope manufacturers tell stories of customers who insisted that the instrument they had bought was no good, even though in reality it might have been of excellent quality. Few people know what to expect to start with and there is no driving test for telescopes! I hope this book will give you a good start and the confidence to know what you are seeing.

There are other books that take things a stage further, either regarding telescopes or astronomy, so I hope that once you have found your feet with this book you will be able to progress further. I recommend Philip S. Harrington's *Star Ware* (John Wiley) for learning more about telescopes and accessories and Norton's *Star Atlas and Reference Handbook*, edited by Ian Ridpath (Pi Press), for more general observing methods. If you would like a basic introduction to astronomy as well as star maps and a constellation guide, modesty does not prevent my mentioning my own *Night Sky Atlas* (Firefly) and the enlarged version, *Complete Guide to Stargazing* (Philip's).

When I wrote the first edition of this book, GoTo telescopes were new on the market. Now they are an established part of the scene, while other instrument designs have been developed. There is more choice than ever before, to the extent that not only the beginner but also more experienced telescope users find it hard to keep up with what's new. So that the book does not go out of date, I have created a dedicated website to accompany the book – www.stargazing.org.uk. There, you can find out about new instruments on the market, more comments on those in this edition, and even let me know your own views.

Because this book is intended for a worldwide market, over a period of time, I've had to give just an indication of prices. That's why you will find references to costs in terms of other purchases – such things as a TV set, hi-fi or car. This may be indirect, but it has its advantages. First, it is independent of your local market conditions. In general, the relative costs of such goods are the same wherever you are. Second, it doesn't date very much. And third, it puts into context the cost of a hobby like astronomy compared with other household goods. It would be even more telling to compare it with the costs of some other leisure items, such as sports goods, but comparatively few people are aware of the price of skis, golf clubs or ice hockey gear, for example.

Actually, you can enjoy astronomy at minimal expense, even if your skies are poor. We tend to see the photos taken by people with large telescopes, expensive equipment and access to ideal observing sites, but you can get a great deal of enjoyment from observing outside your back door with a small telescope or even binoculars. The important thing is to get out there and use your eyes, and to do your best with what's available. It costs very little to make friends with the Universe.

► *The Milky Way in Sagittarius is a very rich area for deep-sky objects. This photograph was taken by Michael Stecker.*

1 • THE STAR PARTY

There are certain places where astronomers and telescopes gather. As the promise of a clear night beckons, cars begin to roll up in some lonely, remote field. People set to work bringing out tripods, tubes, mountings, counterweights, power packs, boxes of eyepieces, stepladders and many other gadgets. As the skies overhead darken, telescopes begin to probe the heavens. No light is seen except for an occasional muted red glow. People pick their way carefully through the darkness, among the instruments, waiting patiently for the chance to peer at a faint blur of light whose photons had started on their journey maybe before there were creatures on the Earth.

This is a star party, where the celebrations are in honor of the cosmos. The food and drink are mostly provided by the heavenly bodies themselves, which the participants devour with their eyes.

A star party is an ideal place to meet astronomers and their telescopes. But few people are really able to spend much time test-driving various different telescopes before they buy, so let me introduce you to some only slightly fictional amateur astronomers and their telescopes, who are attending this particular star party. In this way, we can meet a range of different requirements and instruments, and start to learn some of the jargon of astronomy.

Meet Tom, who is in his late 50s and is a high-ranking local government official. He is a useful guy in the local astronomical society as he helps smooth the way when it comes to getting meeting halls, and he also keeps an eye on planning applications for potential light-pollution problems. He has bought himself a state-of-the-art Meade LX200 telescope with computer control.

This has a short, stubby blue-painted tube only about the size of a large bucket. It sits inside a two-pronged metal fork on a rotating base, atop a shiny metal tripod with fat, solid-looking legs. Every so often there is a squeal from the mounting as the telescope slews across the sky to view a new object.

In his hand, Tom holds a handset with a numerical keypad and various other buttons. To find an object, all he has to do is to punch its name into the handset and off it goes automatically. He demonstrates. "Let's find NGC 4565," he says, feeding the name into the handset. The telescope whirs into action, crawling across the stars until it settles on a spot in the constellation of Coma. It takes about 30 seconds to find its target. Tom peers into the eyepiece, which is at the bottom end of the stubby tube, but has a right-angled attachment to make it easier to view. "Bang on target!" he exclaims after a few seconds. Expectantly, we step forward for a view.

To start with, it's hard even to find the bit we are supposed to be looking through. By feeling the cold outline of the eyepiece against our eye socket, we know we are in roughly the right place. We notice a dim circle of light, speckled with a few stars. This must be it. But what are we supposed to be seeing? We scan the field of view, studying each star in turn. They each seem to have some kind of structure, but it's hard to make it out.

Then, while looking at a star near the edge of the circle of light, we notice something else – a faint, fuzzy spindle near the center of the field of view, looking like a pale gray cigar, fatter in the middle, but barely visible. "See it yet?" asks Tom. We describe what we've seen. "That's what I saw," he agrees. "It must be an edge-on spiral."

An edge-on spiral what, we wonder, but Tom is already up to something else. "I'll increase the power," he says, deftly replacing a black cylindrical object at the eye end of the telescope by another. He stoops again and twiddles a knob on the back of the telescope. "There. Now try." We look again and this time the cigar is somehow easier to see, being a good deal larger. As we stare, we can make out that there is a dark band across the center, in line with the main spindle. "How much does this magnify?" we ask. "It's on 112 now," says Tom, "but it was 77 before I changed magnifications by using different eyepieces. This is my favorite for looking at galaxies – an 18 mm."

▼ Star parties can be huge affairs. The Texas Star Party, shown here, attracts many hundred participants. One amateur observer is carrying out maintenance to his large Dobsonian telescope.

"What is it we're actually looking at?" we want to know. Tom consults a book. "It's an edge-on spiral galaxy about 50 million light years away. So the light we've just seen left it 50 million years ago."

"Will it look the same in a year's time?" we want to know. "Should do," says Tom. "Unless there's a supernova, which is a kind of exploding star. Then you'll see a star in it. Some people spend all their time looking at galaxies hoping to discover one."

"But doesn't it rotate? Will it always be edge on to us?" Somehow we sense that with this question we have shown our ignorance, but Tom pauses only briefly before replying that although it does rotate, it is so distant that it would be millions of years before there is a change.

Not wishing to outstay our welcome, we thank Tom and move on. Within seconds the still night is broken by the whine of the Meade's motor as it goes in search of yet more ancient light.

Maybe we should aim a little lower. Nearby is a much smaller telescope, the sort that we recognize. It has a tube about the size of a piece of small drainpipe, and is on a small wooden tripod. This is a basic 70 mm refractor, with a lens of 70 mm diameter at the top end of the tube – the sort that many people consider for a first telescope. Its owner is a chirpy woman in her 30s, Sheila, who is a school teacher.

Sheila is only too happy to show us what can be done with her telescope. First on the list is Jupiter, which looks like a very bright star high in the sky. To find it in the telescope without the aid of any computer, Sheila first looks through a tiny telescope fixed to the bottom end of the telescope tube and pointing in the same direction. "This finder isn't any good for faint objects," she admits, "but it helps for the bright ones." Soon she peers into the eyepiece at the bottom end of the tube, also at a right-angle like Tom's. "Here we are. You might have to adjust the focus for your eyes because I normally wear glasses. You have to turn this knob to focus." She indicates a shiny knurled knob behind the eyepiece.

We look, and see a small, bright but slightly fuzzy cream-colored disk. As we touch the focus knob, the image shudders violently. We can barely see the image, and have to wait for it to settle down, which takes a few seconds, before we can check the focus. Eventually we have a sharp image and can begin to study it.

The first glimpse is not very impressive. We can see Jupiter's pale disk, which appears quite small in the eyepiece, with what seem to be three stars arranged on either side. But as we look, more becomes visible. Sheila is chattering excitedly about what we can see. "You can see three of Jupiter's moons. There are four big ones but the other one must be behind the planet, or in front of it, so it merges in with the rest of the planet. And you can see the belts very clearly – I can see several white spots but I can't see the Red Spot."

We are staring at Jupiter, trying to make out the detail, when we realize that it's moving leftward, and is now almost out of the field of view. Sheila says that this is because her telescope doesn't have a drive. "Just move the telescope and you can follow it." We do so, and at first it sticks so we push harder. Then the planet shoots off in the opposite direction to what we expected and Sheila has to take over to bring it back into the field of view.

She explains that it is the Earth's rotation which causes objects to move through the sky, which can be overcome using a more expensive motor-driven mounting. "This is just a cheap telescope," she admits. "But it shows quite a lot of detail. Look, Mars is just rising, over that tree. Let me show you." She swings the telescope toward a distinctly pinkish bright object, but she has to report that the view is not wonderful. We look, and see a tiny shimmering pink blob with no structure to speak of. "It's too low," says Sheila. "The seeing's too bad until it gets higher. Come back later."

Next on the list is an extraordinary device, which if we did not know was a telescope would be hard to put a name to. Mounted in what

▼ *Face-on spiral galaxy M101 as seen through a 450 mm home-built Dobsonian, drawn in "negative" by Martin Lewis.*

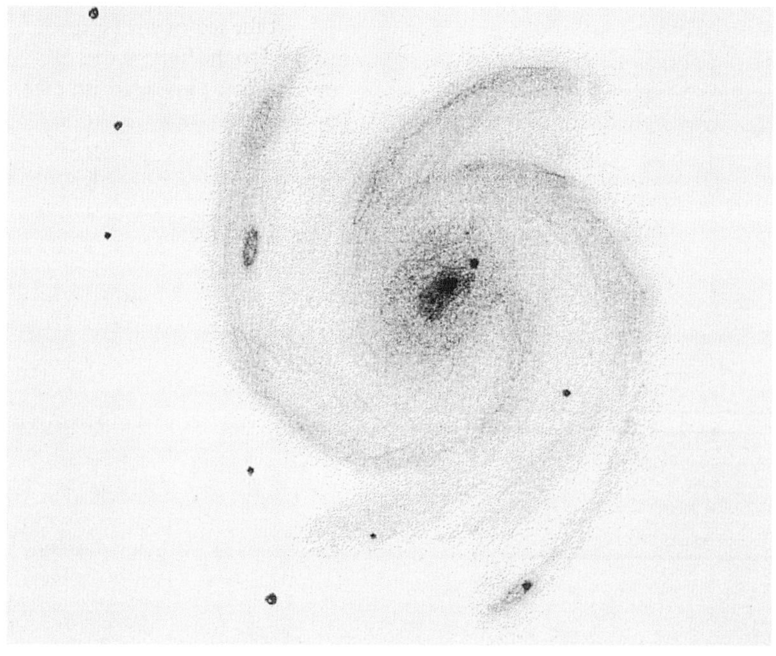

seems to be a wooden box is a large cylinder of thin black plastic sheeting, which appears from its shape to be surrounding a framework of rods and hoops. Although the cylinder is pointing skyward, its user, a man in his 30s, is perched on a short stepladder looking into an eyepiece which projects sideways from the top end.

This is Phil, an engineer by profession, who has built this telescope himself. He calls it his "Dobsonian light bucket," and he says it is an 18-inch telescope. "That's 450 mm – it's got four times the light-collecting area of that LX200, and it cost me a quarter of the price," he boasts. "And," he adds, "it's silent!"

We beg a look, which he happily allows. "This is M101, a face-on spiral galaxy," he says. We climb the stepladder, trying not to grab hold of the telescope to steady ourselves, and peer into the chunky eyepiece. Compared with the 70 mm refractor, the field of view is vast. A scattering of stars fills the image, and in the center is an unmistakable glow of light. It is not brilliant, but by looking around the image we can make out a spiral structure, with a rather brighter knot of light at its center. The image remains steady but very slowly drifts off to one side. Phil tells us to push the telescope to track it. Gingerly we do so, and are amazed after our experience with the 70 mm to find that it glides easily, then stops with hardly a judder. "Is it on ball bearings?" we ask. "Nope – just Teflon pads," says Phil. "That's the secret of the Dob. Cheap, low-friction surfaces, no complicated bearings. And it's easy to

▼ *Most star parties are quite small gatherings. Here, a 100 mm refractor is being readied for the night's observing.*

make, though if you want a portable instrument like mine it helps to be good at making things. I can get it out of the car and be observing in five minutes."

Impressed, we wonder what the same object looks like through Tom's LX200. He obligingly commands the instrument to find it, and we see the same object but rather fainter and paler. The spiral structure is not so obvious – if indeed we are seeing it at all, rather than just interpreting the hazy glow as a spiral because we know it to be so.

Our next view is through a neat-looking telescope on a metal tripod, rather fatter than the 70 mm refractor and with its eyepiece at the top. The man looking through it is in his late 40s, who turns out to be Pete, a plumber. He explains that this Celestron NexStar 130 is a very portable computerized reflecting telescope. It has the same basic optical design as the big Dobsonian, but with a mirror just 130 mm or 5 inches across, rather than 450 mm. Even so, it's big enough to give pretty views of the bright planets and will seek out thousands of fainter objects at the press of a button.

"I was amazed how easy it is to set up," he says. "I just point it at any three stars in the sky, even if I don't know their names, and then it knows where to find everything else." Jupiter through the NexStar is considerably larger and brighter than we remember it from Sheila's 70 mm refractor. Now we can see the darker, slightly brownish belts and can even make out some of the spots that she was enthusing about. But one change is obvious – there are now four moons, instead of just three. The new one is just a short way from the edge of the planet, more or less in a line with the others. "You can see its shadow on the planet," says Pete. Sure enough, there is a small black dot on the planet's disk, slightly offset from the line of the satellites. By now we have become used to some of the tricks of observing – getting our eye in the right place, altering the focus to suit our eyes, not bumping into the eyepiece, though when we do the NexStar takes a few seconds to settle down.

But we still want to see Mars. Is this "seeing" any better? The planet is now rather higher than before, and Pete finds it for us. The disk is still tiny, but we can definitely make out that it is a disk and not just a blob. The whole thing is by no means steady, even though the telescope has a motor drive. It ripples and swims around slightly, almost like some single-celled creature seen through a microscope. "That's bad seeing for you," says Pete. "We very rarely get a rock-steady view when the planet is low – it's our own atmosphere that causes the trouble."

As we watch, Mars goes through its jig, sometimes appearing as three or four overlapping disks. How can anyone see any detail on such a capricious object? Then, suddenly, just for an instant, the planet settles down and we glimpse the white gleam of a polar cap and a darker smudge near the middle of the planet. Pete shows us a drawing he has just made, while sitting at his little telescope on the picnic table. Inside a circle a couple of centimeters across he has skilfully pencilled in the dark marking and some others, and shown the polar cap by a dark line surrounding it.

Also in his book are some he has made on previous nights, which he has worked on with some artistic skill. He has added a black background, and shown a mass of detail which we are amazed by. "You just have to wait for the good moments," he explains. "The amount of detail you can see is usually limited by the seeing, rather than the telescope. Even a smallish telescope can show a lot of detail if the air is steady enough. Tonight we've got good transparency but bad seeing. The air is clear but unsteady, but when it's steady it's often more misty."

Meanwhile, over in another corner of the star party, some way from the others, is a figure hunched over a glowing laptop screen. He is watching the monitor intently, as it shows a tiny star jiggling around in the fashion that we now realize is caused by the "seeing." A green circle surrounds the star, faithfully following all its jiggles. Presently the laptop screen displays an image of stars. We find out that this is Andrew, who works in a bank by day.

He tells us that he is using a 15 cm reflecting telescope to photograph a pair of galaxies in Leo, known by their catalog numbers as M65 and M66. This requires exposure times of several minutes on his digital camera, and means that the camera has to follow the movement of the sky and the vagaries of the seeing absolutely accurately.

"I've got a Vixen Sphinx mount," he explains. "It tracks the sky very precisely, once I've lined it up using the built-in pole finder." He shows us a small telescope at the lower end of one of the axes of the blocky white mounting. "But even the Sphinx can't allow for everything, particularly the seeing. When I started out doing this I had to spend all my time watching for any drift of a guide star, and correct it instantly. But now technology does the job for me."

He points out a smaller refracting telescope on top of his reflector. Instead of an eyepiece there is a tiny electronic gadget. "It's a webcam, the sort people put on their computers to send their image with Internet phone calls," he explains. "I point it at a star near the object I want to photograph, and it does all the tracking for me automatically." We look through the camera viewfinder itself, but cannot see any galaxies. Andrew then shows us the image he has just taken on the laptop, and the galaxies stand out clearly as fuzzy ovals. "They'll show up much better when I've processed the image," he says. "But at least I know I've got the shot. When I used film instead of the digital SLR, I could waste a whole night taking pictures without realizing

► *Observing with a 600 mm Dobsonian telescope. The eyepiece is some way off the ground, so a stepladder is essential.*

that the camera wasn't properly in focus, and not find out until I got the film processed!"

The telescopes we have seen at this star party are just a sample of the wide range available, and vary in price from the cost of a simple autofocus camera up to that of a reasonable used car. The people using them are also representative of the wide range of amateur astronomers, who come from all walks of life.

They may seem knowledgeable about their subject, and about the telescopes that reveal the Universe to them, but they all had to start somewhere. With astronomy, as with anything, what seems baffling at first becomes simple with a little experience. I hope that with this book to guide you, you will have a head start.

2 • WHAT'S AVAILABLE

There is no perfect telescope. Each one, as long as it is of reasonable quality, will show you fascinating things, although in general the more you pay the more you'll see. Many people have more than one telescope, just as a golfer has more than one golf club. Hopefully, as you read on you will be able to decide which is the right one for you.

There really are dozens of options, and you can easily end up with something that you're disappointed with – and end up not using at all. It all makes choosing a car, say, seem easy by comparison. At least you can test-drive a car and you have an idea of what you want from it.

In the case of a car, though, you probably know before you begin what sort of car you want – a humble hatchback, a swanky saloon or a funky four-wheel drive. Just as each type of car has its advantages and drawbacks, so the different types of telescope have their own uses. To start with, here's a nutshell guide to what the types are. Following that, we go into more detail about each type and what to look for.

Binoculars

There is hardly an amateur astronomer in the world who doesn't have binoculars, which are basically two low-power telescopes side by side. They have a multitude of uses, from stargazing to serious observing. The advantages of binoculars are that they are good value since they are made for a mass market and they provide a significant step up from observing with the naked eye. Even though they magnify just a few times, they bring into view many thousands of stars and other objects that are invisible with the naked eye alone.

▲ *A selection of binoculars from the Celestron range. Left to right: 15 × 70 with Porro prisms; 8 × 42 with roof prisms; 20 × 80 with Porro prisms; UpClose 10 × 50 with Porro prisms.*

The advice to budding astronomers is often to get binoculars before even thinking about a telescope, though some people argue that you won't be happy until you have the telescope anyway. But it is far better to get cheap but adequate binoculars than a poor telescope costing several times as much – you will get far more use out of them, and they may well last you a lifetime.

Much has been written about the best choice of binoculars for astronomy, but you can't go far wrong with a fairly basic model, such as the classic 7 × 50 or 10 × 50 models, though more lightweight instruments such as 8 × 30s or 8 × 40s are perfectly acceptable. The first figure in each case is the magnification, and the second is the size of the main lens in millimeters. Small telescopes are described in the same way.

Refractors

These are the classic telescopes, the sort you expect to see standing imposingly in a bay window. The cheapest telescopes are of this sort, such as little hand-held ones and those in camera-store windows.

They have a lens at the top end of the tube and you look up through an eyepiece at the bottom. The smallest are of little use for astronomy, but with a good 60 mm refractor – that is, where the main lens is 60 mm or 2.4 inches across – you can see some details on the planets, get good views of the Moon, and find hundreds of deep-sky objects such as star clusters, nebulae (gas clouds) and some galaxies. The views won't be wonderful, but there's a lot of fun to be had. I say "a good 60 mm refractor," but sadly a lot of small refractors are rubbish. Price is not necessarily a good guide, either, so don't buy rashly.

When you get to larger sizes, however, such as 150 mm refractors, you are in the area of serious telescopes, the sort that many amateur astronomers dream of. Their prices could set you back as much as a brand-new car.

Refractors are noted for their sharp views of good contrast and their lack of maintenance. But their images are usually affected by false color – unwanted colored fringes around objects. False color should be unobtrusive in a well-made refractor, though only the most expensive models are completely free from some trace of it.

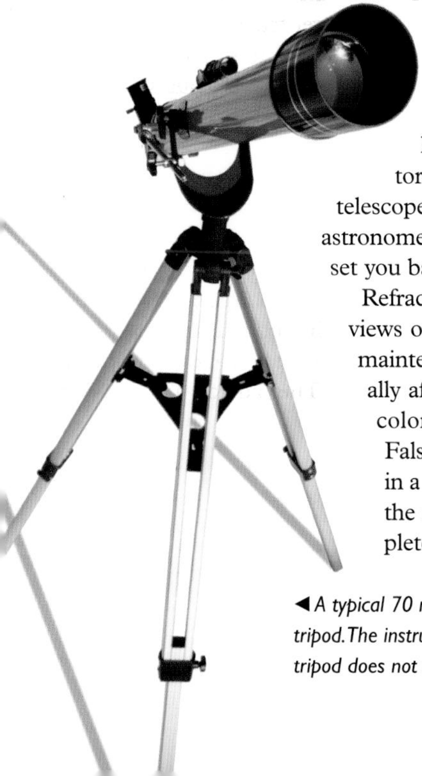

◀ *A typical 70 mm refractor on a basic aluminum tripod. The instrument gives good images, but the tripod does not hold the telescope very rigidly.*

Reflectors

Many amateurs prefer reflecting telescopes, mostly of the Newtonian type. These use a mirror to collect the light instead of a lens, and are free from false color. Once you get above the very smallest models they are much cheaper than refractors, size for size, but they do take some looking after. They are more suited to the committed amateur astronomer than the beginner, as a medium-sized reflector can be quite bulky.

Reflectors are particularly appropriate where light-gathering power is needed, such as for observing faint objects, and all large professional telescopes are reflectors of one sort or another.

Catadioptrics

A third design of telescope combines mirrors and lenses to help overcome this drawback of size. The technical term for such instruments is "catadioptric," and the most common types are the Schmidt-Cassegrain telescope, or SCT, and the Maksutov, also known as the Maksutov-Cassegrain. These have short, stubby tubes that are easier to mount, size for size, than full-sized Newtonian reflectors. The penalties for the compactness are slightly poorer optical performance than a Newtonian of the same size, and higher price.

Catadioptric telescopes have become very popular in recent years, as their compact optical designs are very convenient.

▲ *Observing with a 200 mm Celestron Schmidt-Cassegrain telescope. The unit is motorized, and the compact tube provides eye-level viewing.*

More about binoculars

This book would not be complete without something about types of binoculars. There is no need to spend a fortune on them, though some general-purpose binoculars can cost as much as quite large telescopes. I don't see the point in spending such sums for general astronomy, where weatherproofing and ultra-rugged construction are not really called for. Good basic binoculars are not expensive, though you should avoid the very cheapest. Expect to pay about the same price as for a reasonable compact camera.

For general astronomical purposes, it is a mistake to aim for a high magnification, above 12, say, which will be hard to hold still and gives little advantage. These days, some binoculars have image-stabilization systems that can be very effective, though I would always advise you to try them before buying. Some merely damp out movements rather than actively compensating for wobble, and they are of limited value to the astronomer, though they may be well suited to use aboard ship, say.

Experienced observer David Frydman says that he can see stars a magnitude fainter when using his Canon 15 × 50 stabilized binoculars. Non-stabilized binoculars are not normally made in this size with such a high magnification, and it is only because of the stabilizers that it is possible to use them easily. They do, however, show considerably fainter stars than you might see with cheaper hand-held binoculars for two reasons. One, the stabilized image allows you to spot faint stars that might otherwise move around too much, and two, the instruments must be made to a higher optical specification anyway, as the images are more likely to be scrutinized carefully.

One important factor to consider when choosing binoculars for astronomy is the actual field of view that they give, ideally measured in degrees – in other words, how much of the sky you will see at a single glance. This figure is often either hard to find, not given in degrees or not quoted at all. It is important because many astronomers use binoculars as wide-field finder instruments, to locate objects in the sky that they wish to observe. The smaller the field of view, the less useful they are for this purpose. A field greater than 5° is desirable, which is equivalent to 262 feet at 1000 yards. As the magnification increases, so the field of view inevitably goes down.

There is another type of field of view to consider, however, and that is the apparent field of view of the eyepieces, which is the spread of the circle of light you see when you look through. In the cheaper instruments this is quite narrow, like looking through a tunnel, and is probably around 40° or so. Instruments referred to as "wide angle," often with "WA" in their specification, may have an apparent field of 60° or more. It is possible to have binoculars that despite having fairly high magnifications still yield a good actual field of view because they have wide-angle lenses.

The apparent field of view is rarely quoted in manufacturers' specifications, however. I have come across some binoculars that achieved a wide apparent field of view at the cost of giving a curved focal plane – which means that not all of the field of view comes to the same focus position. This may not matter too much by day, where objects are often at different distances anyway, but when looking at stars it can mean that anything not at the center of the field of view is not in focus and appears like a comet. Such instruments are of limited use for astronomy, though they may be fine for birdwatching.

Variable-power or zoom binoculars are not recommended for astronomy, as the extra lenses needed to change the magnification reduce the optical performance.

Some amateur astronomers favor large binoculars, such as 11 × 80s,

Adjusting binoculars

Step 1. Vary the width between the two eyepieces to suit your eyes. The scale on the pivot shows the separation – here 60 mm.

Step 2. Focus on a chosen object with the center wheel using your left eye only.

Step 3. Finally, focus on the same object with the diopter adjuster on the right eyepiece using your right eye only.

but these are not beginners' instruments. Your first choice should be an instrument that is easy to use without requiring extra support. The bulk of binoculars depends entirely on the size of the main lenses, not the magnification. For prolonged use, even the traditional 50 mm binoculars can get a bit much, so you may find that models with 30 mm or 40 mm lenses, while not giving as bright an image for the same magnification, are better.

There are two alternative designs of binoculars as a result of different ways of condensing the light beam. Without a folded beam, binoculars would be long and unwieldy. The classic style uses a pair of what are called Porro prisms to fold and invert the light beam. You can recognize these because the eyepieces are not in line with the objectives, which have a wider separation. Another way to fold the beam is by means of roof prisms, which allow the eyepieces to be in line with the objectives. These are more compact, but there is no particular weight advantage.

Points to consider when choosing binoculars are the optical quality, the apparent field of view, the lack of distortion and the flatness of the field of focus. If you wear glasses, you may want to be able to observe without taking them off. This is more than personal preference, because if your glasses correct for astigmatism you will not get a good view without them. Often, the binoculars have rubber eyecups that fold flat so you can press your glasses up against them. They should also have good eye relief, which is the distance from the eyepiece that still gives a full field of view. But if your glasses simply correct for long or short sight, the focusing system of the binoculars will usually compensate unless you need very strong correction.

Many people do not know how to adjust binoculars. The method is first to adjust for the separation between your eyes. There may be a scale engraved on the pivot, which is the separation in millimeters. If you are not looking directly down the center of each half of the binoculars, you will get a distorted view no matter how good they are. So make a note of your setting, which should apply to any binoculars.

Next, use the left-hand half only to focus on a distant object using the center focusing wheel. Then use the right-hand half only, and focus on the same object using the diopter adjuster on the right-hand eyepiece, without changing the focus of the center wheel. Again, there may be a scale, which should be the same on all binoculars, so note this setting as well. Now you can pick up any binoculars and set them to your eyes instantly. Some binoculars do not have a central focusing wheel, however, and the two eyepieces must be focused separately.

You will probably not be able to test binoculars on the sky before buying, but do insist on trying them on a fairly distant object,

How to test binoculars by day

• Having adjusted the binoculars as described on page 18, you should be able to see the entire field of view without discomfort, such as the eyepieces pinching your nose – a problem with some wide-field designs.

• Choose a high-contrast test object. A thin wire or antenna against a bright sky (but not dangerously close to the Sun!) is ideal. Look at it carefully for signs of false color at the edges of the object, which should be minimal. Also look at the contrast between the dark object and the bright background. There should be no hint of glare, and the object should be distinct with no blurring. There should also be no obvious double vision, which is caused by the two optical systems not being parallel to each other. Sometimes your eyes are able to pull two slightly separate images together, which you may not notice at first. But when observing over a period of time this can become uncomfortable, so check carefully if your eyes seem to feel strained.

• Now move the binoculars so that the object is close to the edge of the field of view. The performance will almost certainly not be as good, but in particular you should not have to refocus. Only at the very edge should the distortions be noticeable.

▲ *The view through the objective.*

• Scan the binoculars across a scene, particularly one with a repeating pattern such as a brick wall. Check that the image does not seem to swell or shrink at the center, which implies a change of magnification across the field of view.

• Check each model against others of the same specification. Points to look for are the width of the field of view, brightness of the image and any overall color in the image. The sheen on the front lenses, sometimes a startling red color, is called blooming or multicoating, and is designed to improve the light transmission. It should not affect the image color. Try to look through more expensive models to see if you can tell the difference.

• Look and feel for mechanical problems – loose or stiff movements and, in the case of Porro prism binoculars, play in the arms that carry the two eyepiece assemblies. These should move smoothly and without sticking as you focus.

• Finally, look into the objective lenses so that you can see the circle of light at the far end coming in through the eyepiece. Look through the edges of the objective. The internal components should be clean and free from defects – particularly important if you are buying secondhand. And if the eyepieces are partially obscured by the internal prisms at the edge, you may be losing light. This depends on what is called the exit pupil – the size of the beam of light emerging from the eyepiece. You calculate this by dividing the objective size by the magnification, so 10 × 50s have an exit pupil of 5 mm and 7 × 50s have an exit pupil of 7 mm. So the full central 7 mm of the eyepiece of 7 × 50 binoculars should be visible when looking through the extreme edge of the objective.

ideally outdoors. Do not test through an open window, because the temperature difference between indoors and outdoors can produce air turbulence that can spoil the image. A closed window can also introduce distortions. But bear in mind that thermal effects such as heat haze can make distant objects at ground level shimmer, in which case aim for an object high up and not too distant.

Terrestrial telescopes

The smallest telescopes are usually terrestrial refractors – that is, they include as part of their design optics that give an erect image for everyday use. There may be interchangeable eyepieces for giving different magnifications, or there may be a built-in single or variable-power (zoom) eyepiece. An astronomical telescope, by comparison, does not give an erect image, and would need additional optics (either a lens or a prism arrangement) to do so, as well as the usual range of eyepieces.

What's wrong with using a terrestrial telescope for astronomy? The additional optics needed to erect the image inevitably absorb some light, and may introduce distortions that become only too evident at night, though they may be negligible by day. These distortions should not be troublesome in a well-made instrument, but in the cheaper models the astronomical performance could well be compromised.

If you do want a small and portable telescope for everyday use, with a power greater than that provided by binoculars, there is quite a range available of what are called spotting or field scopes, which are used by bird enthusiasts and rifle shooters in particular. These range widely in specification and price, but most are rather expensive for the optical performance that they offer because they are designed for hard use in the field rather than for careful use in the garden. However, someone who wants a small telescope primarily for general-purpose daytime use with the option for some occasional night viewing might prefer a terrestrial telescope.

The availability of small hand-held telescopes other than spotting scopes is limited. Ignoring small brass telescopes that are made for looks rather than for performance, it can be difficult to find good-quality but reasonably priced small telescopes. A straightforward telescope would have a fixed magnification, such as a 20 × 50 or 30 × 60. These combine a fairly bright image with a magnification that will show some detail on the Moon and Jupiter. If they are optically sound, they should also show you the brighter nebulae and clusters. Some Russian-made telescopes of this specification are of good quality, but others are poor. Availability changes from time to time, so check www.stargazing.org.uk for the latest information on such instruments.

What to avoid in a small telescope

- False color around the edges of high-contrast objects, such as wires seen against a bright sky (not anywhere near the Sun for safety!).

- Changes in the shape of point-like objects, such as holes in guttering or decorative ironwork, on either side of the focus position. This suggests a fault known as astigmatism.

- Generally low-contrast images. About your only guide here is to look through a top-of-the-range instrument, hoping that it is as good as its price suggests, for a comparison.

- Flaring or paling of the whole image when there is a bright object such as a light bulb in the field of view. This indicates poor baffling of the inside of the tube. To check for this, remove the eyepiece if you can do so and see if the insides of the tube are shiny, so that they reflect the light, or are well baffled so that the only light you see comes from the objective lens itself, as should be the case.

◀ *False color as seen through a small non-achromatic telescope. Although the image is fairly sharp, the blue and red fringes are very obvious.*

There are also small zoom telescopes available, with specifications such as 15–45 × 50, but they are variable in quality. Cheap zoom telescopes are not generally recommended for astronomy because the zoom system requires additional lenses that introduce distortions unless they are very well made. I have seen some that give reasonable images at night, but in general they are not good.

Some of these terrestrial telescopes are sold by themselves, usually with a tripod bush so that they can be fixed to a standard photographic tripod. Others have table tripods with short legs, and look virtually the same as the smallest astronomical telescopes. Many of them really come into the category of toys, and should be avoided (see "How to avoid a lemon," page 33).

More expensive birdwatching or spotting scopes have interchangeable eyepieces and may take camera adapters. For the same price, however, you can get a much more powerful astronomical telescope, so they are only worth considering if you need them primarily for their portability and ruggedness as birdwatching scopes. These instruments are typically priced at about the same as a good SLR camera.

Shop-testing small telescopes

Unfortunately, it's not easy to test a telescope by day for its suitability for astronomy. The reason is that astronomical objects are severe tests of optics – they have very high contrast compared with most daytime subjects. To use the hi-fi analogy, testing optics by day is like trying to listen to a hi-fi unit in a department store with dozens of other competing sounds going on around you. Get the unit home and listen to it with no distractions and you might spot distortions and rattles that you missed in the store.

A star image is the equivalent of a single, pure note. Seen by itself at night, you can see any slight imperfections or false color. A daytime view, however, is like viewing millions of single points of light of different colors and brightnesses, all cheek by jowl. Any imperfections can easily be masked. Bear this in mind when you are choosing any optical instrument, whether a telescope or binoculars.

In general, the tests suggested for binoculars (see page 20) can also be applied to terrestrial telescopes, with the additional opportunity of increasing the magnification. But bear in mind the other difficulties of daytime testing. You will probably have an unsuitable location, such as viewing through the open door of the shop or outside in the street. It can be hard to use anything like the same magnification that you would under ideal conditions at night, when atmospheric turbulence has hopefully settled down and the telescope is at the ambient air temperature.

However, a poor telescope that shows its true colors by day will be hopeless at night, so at least you can rule out the real bad 'uns.

Finally, unless you are buying from a telescope specialist, be very wary of any assurances from the sales staff. Only an experienced observer can really say how good a telescope is, and even then, often only with a known good instrument available for comparison.

Small refractors

Virtually every camera store and shopping catalog has a small refractor available, usually a 60 mm or 70 mm. They are also to be found in some toy shops and the toy departments of large stores. For most people, indeed, these are about the only places where they will see a telescope and, not surprisingly, they are often the first port of call when they want to buy one. The leading popular telescope manufacturers also have them on their lists.

So, how good are small refractors, and are they a suitable purchase for beginners? And can you trust these sources of supply? It would be nice to be able to give a straightforward answer to each of these questions, but in practice the situation is not simple.

There's no denying the appeal of the small refractor. The 50 mm or

60 mm models come in handy cardboard boxes that you can keep under the bed or wherever. They are made in large numbers so they're comparatively cheap. They will lie around unused for years if necessary, as long as they are kept dry, and still be ready for use at any time. Unless they are sadly mistreated, they need little maintenance. And above all, they can show pretty views of some of the wonders of the night sky.

With even a 50 mm refractor you can start to get acquainted with a whole range of astronomical objects. The bright planets become distinct worlds, and you can study literally hundreds of deep-sky objects, including double stars, nebulae, star clusters and galaxies.

Increase the aperture to 60 mm, and the views become brighter and more distinct. Usually the difference in price between a 50 mm and a 60 mm instrument is not very great, so it is better to go for the larger aperture if you can. Having said that, a good 50 mm refractor is better than a poor 60 mm one – a general rule that extends to all sizes and types of telescope.

The standard 70 mm refractor

This is such a common telescope that I'll spend some time talking about how it works and what you can do with it. At one time, the smallest telescope in every manufacturer's range was a 60 mm, but in the past few years this has moved upward to a 70 mm, once rare, and there are fewer 60 mm scopes to be seen. The extra 10 mm of aperture compared with a 60 mm brings in over a third more light, so faint objects are that much easier to find.

Actually, there is no such thing as a standard 70 mm refractor, although many of them appear superficially the same. A cheap model will have a plastic tube and fittings and may have low-grade optics. At the other end of the scale its more expensive cousin will have a well-made metal tube and fittings, and optics that perform as well as a telescope of that size can be expected to. There could be a factor of three or more between the prices of the two, but seen side by side in a camera store window you might have to look quite carefully to tell them apart.

Even two instruments that look identical may perform quite differently. Virtually all these telescopes are now made in the Far East, and even those bearing well-known and respected Japanese trade names may be made in Taiwan or China. Subcontracting is standard practice, so the same parts may be found in telescopes bearing different names and at widely different prices.

However, making the optics of a telescope is not a simple process like making a moulding. The final finish is very critical, and two lenses of the same specification from the same plant can turn out to be very different. They could be put into telescopes with many identical parts, but the performance could be excellent from one and terrible from the other.

Quite often, the more expensive telescopes will also have a longer focal length. Many 70 mm refractors have a focal length of 700 mm, giving them a focal ratio of f/10 (see page 29 for a description of focal ratio). Others have a focal length of 900 mm (f/12.9). Curiously, it is easier and cheaper to make a lens of long focal ratio than short, yet the images it gives could well be more pleasing. However, manufacturers often charge more for the longer focal length instruments. If you have the choice of a short- or a long-focus 70 mm refractor, go for the longer one every time if you want to observe details on planets.

One perennial problem with small refractors is the flimsiness of their mountings. One touch and the whole telescope vibrates, making focusing awkward. The lack of sturdiness also makes it hard to change eyepieces to give a higher magnification, as the telescope often shifts slightly in the process. Instruments with plastic tubes are particularly prone to this, while well-made instruments may often cost considerably more as a result.

The advertising and specification of small telescopes is often designed to catch out the unwary. The classic ploy is to claim unrealistically high magnifications for the telescopes. The box on page 34 tells you how to work out the magnification given by a particular eyepiece. Many telescopes include what is called a Barlow lens, which doubles or triples the magnification of any eyepiece. A standard kit may therefore include a 20 mm and a 12.5 mm eyepiece, with a 3× Barlow. With 700 mm focal length, a 20 mm eyepiece gives a power of 35 and a 12.5 mm a power of 56. Add a 3× Barlow and you get powers of 105 and 168. Some small telescopes claim to give magnifications as high as 525, which sound impressive but in practice are pretty useless.

◀ *A small refractor is often the first telescope for a child. If it is well made, it will show details on the Moon and planets and could be the start of a lifetime's interest.*

▶ *Two 70 mm refractors: at left, the Sky-Watcher Mercury 705, 500 mm focal length (f/7.14) on an AZ3 altazimuth mount with slow motions. At right is the cheaper Mercury 707, 700 mm focal length (f/10), on a more basic AZ2 mount.*

There is a practical limit to the amount of magnification that a small telescope can give. This is set by the resolving power of the optics (see page 36). It's generally accepted that the highest power you can use on any telescope is twice the aperture in millimeters – and that requires excellent conditions and good optics. So a 70 mm refractor has a natural limit of 140 on its magnification. What's more, if the optics are not perfect, using high powers will give poor results anyway. Quite often, the Barlow lenses themselves are non-achromatic, so the image is garishly false colored. The magnification may well be as described, but it is unusable.

I have heard it said that high magnifications sell telescopes, though I'm sure that if people were asked they would prefer a telescope that performs well, within stated limits. Regrettably, few people really know what they are buying and so the manufacturers get away with it every time. In fact, many perfectly good telescopes provide these improbable magnifications, so you can't guess the quality based on that alone.

What can you expect to see through a good 70 mm refractor?

On the Moon, you will be able to see a vast amount of detail – with stunning views of craters, mountains and seas. This is not to say that a larger telescope will not show more, but the view will be seriously impressive.

When it comes to the planets, the results will vary considerably with the size of the planet at the time. Jupiter is easy to view whenever it is visible, and in addition to the main belts and satellites you should be able to make out some detail in the belts, such as any reasonably large spots, though any such details will probably be at the limit of your perception rather than easily visible.

Saturn's rings are equally easy to see and a darker belt on the planet's globe may be visible.

Venus can also be a large planet in the sky, although it varies in size depending on its distance from Earth. The crescent phase, seen when

How a refractor works

The basic optical system of a refractor is very straightforward. A lens at the top end of the tube refracts (bends) light to form an image at the bottom end. This image can be seen on a screen or photographed by putting film there, but the most common method of observing is to use an eyepiece to magnify it. This gives an upside-down image, which is not usually a problem for astronomers.

This viewing position, looking directly through the tube, requires either crouching on the ground or a very tall tripod. An alternative is to use a star diagonal consisting of a right-angled prism that reflects the light through 90°, providing a more comfortable viewing position but at the cost of introducing extra glass surfaces and laterally reversing the view.

It may seem odd that a prism, noted for splitting light into its component colors, is used to reflect light, but if the light passes through the faces almost at right angles it is reflected off the back face with virtually no false color. The advantage of this over a mirror is that it gives a bright reflected view with no coating that can tarnish.

A simple lens produces bad false color, so all lenses in practical telescopes are achromatic, consisting of at least two components made from different types of glass so the color dispersion of one cancels out that of the other to some extent. But this is only completely true at one selected wavelength, so there is inevitable color fringing, usually blue, in the images given by basic achromatic lenses.

it is closest to Earth, is a beautiful sight, although little or no detail is likely to be visible on Venus.

In the case of Mars, only when the planet is at its closest, which happens roughly every two years, will a small refractor show any detail, and then only a hint of a dark marking and a polar cap. As for Mercury, Uranus and Neptune, be happy just to see them, and don't expect anything more than a dot of light.

On deep-sky objects, a 70 mm refractor will actually show hundreds of objects, given good conditions, but your view of many of them will be limited to a small fuzz with no hint of structure. Don't expect to see spiral structure in galaxies, for example. But the brighter nebulae and star clusters will be visible, and certainly all of the Messier catalog (see page 136) given the right circumstances. The word, however, is visible rather than spectacular.

Briefly, then, a basic 70 mm refractor will satisfy your curiosity to

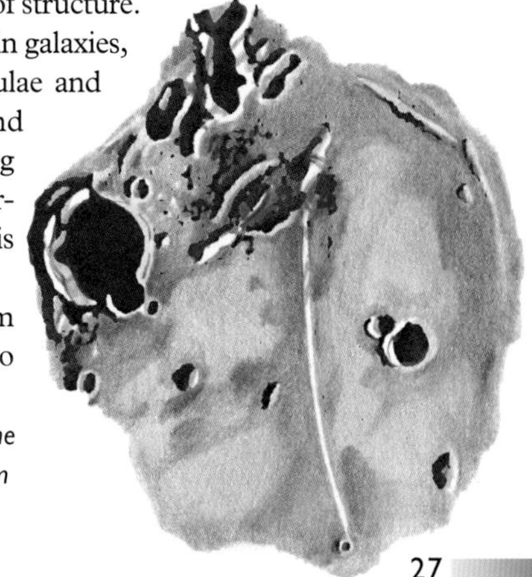

► *A drawing of the Straight Wall area of the Moon made by Peter Grego using a 60 mm refractor with a magnification of 100.*

see for yourself the most popular objects in the sky. But what it won't do is to show them very well, and the limitations of its size and mounting will not allow you to photograph most of them satisfactorily. No matter how well it is made, a 60 mm or 70 mm refractor has its limitations and if all you want to do is observe, go for aperture rather than finish, with the provisos that the optical quality of the lens should be good and the mounting should be adequate to the task.

Where small telescopes are concerned, what you get for your money is quite different from some other consumer products. For the price of a reasonably good 70 mm refractor on altazimuth mount, you can buy a perfectly acceptable DVD recorder or digital TV recorder, packed with microelectronics and precision engineering. Yet the 70 mm employs no more than 18th-century technology, and the parts could be made in a small workshop using simple tools. The difference is that virtually every home has at least one TV recorder, while maybe one in a thousand has a telescope. Even with a global market, telescopes can't be produced with the same economy of scale as TV recorders.

Larger refractors

At one time, before the big telescope manufacturers came on the scene, the 3-inch (76 mm) refractor was regarded as a fairly serious amateur instrument. It was the minimum size with which you could start to undertake observing programs. By today's standards, however, it is really rather small, even if you upsize it to the 80 mm mark. Even so, this is an extremely popular size of telescope, and not just among beginners. Many advanced amateurs will have an 80 mm refractor in their collection. For the beginner, however, this is an affordable size of instrument that will allow them to see a bit more than the smallest refractors. You might wonder whether an extra 10 mm or so of aperture would make a difference, but bear in mind that when it comes to seeing faint objects, it's the light-collecting area that counts. This increases with the square of the aperture – the area of a circle is "pi-r-squared," remember? So while a 70 mm telescope has an area of 4900 square millimeters, an 80 mm aperture has an area of 6400 square millimeters – 30 per cent larger.

A good 80 mm or similar aperture will start to show you significant amounts of detail on the Moon and planets. For example, with a 70 mm telescope you cannot see what is called the Cassini Division between the two brightest rings of Saturn. However, an 80 mm refractor should just show it. The Cassini Division is not exciting in itself, but it is a fair test of the detail visible. See the box on resolving power – "How much can you see?" – on page 36 for more details.

When it comes to deep-sky objects, an 80 mm refractor brings in a

Focal length and focal ratio

Every lens and mirror has a focal length and a focal ratio, not to be confused and very important for all sorts of reasons.

Take a magnifying glass and hold it so that it forms an image on a wall of a bright, distant object such as a streetlight. You could use the Sun, but you run the risk of scorching a mark on the wall! The distance from the magnifying glass to the wall is its focal length (strictly only true if the streetlight is at infinity, but good enough to get the idea).

If you had another magnifying glass of the same diameter, but thinner at the middle, it would also cast an image on the wall but you would need to hold it farther from the wall to get the image. So its focal length is that much longer, and you would notice that the image it produced was bigger but dimmer. The same scene is spread over a wider area, so if you were to try and use the lens in a camera, say, you would get a more wide-angle view with the shorter focal length than with the longer, for a given size of film. Notice that the diameter of the lens has nothing to do with its focal length.

You can see that the longer the focal length, the higher the basic magnification of the lens. Concave mirrors of the sort used in telescopes work in the same way. And the same principles apply to telescopes based on such lenses or mirrors: the longer the focal length for a given aperture, the larger the image it will give but the dimmer the basic image will be. You often have to decide, when choosing a telescope, whether you want one that gives a bright image with low magnification or a dimmer image with higher magnification.

Actually, in the case of a telescope you use an eyepiece as well, so you could get the same magnification from both lenses by using different eyepieces and much of the time the view would be virtually identical in each case. So what's the point of having a choice of focal lengths? For a given aperture, a short focal length will be compact and will give comparatively wide-angle views of the sky, particularly useful if you want to view or photograph some deep-sky objects and comets. But it will not perform as well at top magnification as a telescope of longer focal length, that will be more suited to viewing and photographing the planets.

Sometimes a telescope's focal length in mm is quoted, but more usually the focal ratio is given. This is the length divided by the aperture, so a telescope of 100 mm aperture and 1000 mm focal length is f/10, while if it is 500 mm focal length it is f/5. People may talk about the shorter focal length being "faster," as if it were a photographic lens. If you want to photograph an extended object, such as a comet, the smaller f-number will mean shorter exposures. For ordinary reflectors and refractors, a small f-number such as f/4 means a stubby tube, while a large f-number, such as f/10, means a long one. But in the case of Schmidt-Cassegrains, even an f/10 instrument has a stubby tube because the light path is folded up within the tube.

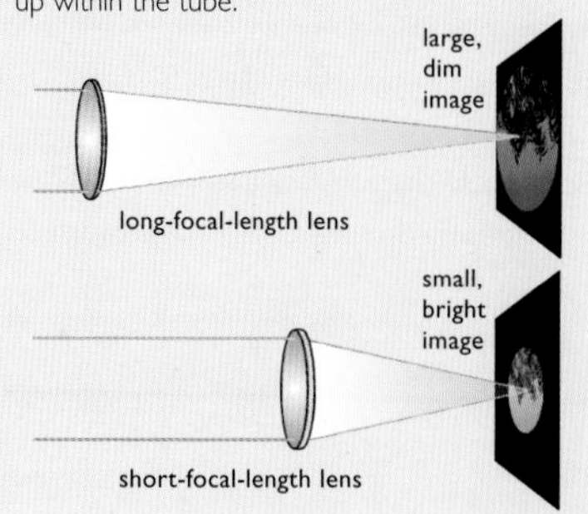

► *Lenses of the same diameter but different focal lengths, and the images they produce. The long-focal-length lens has less steep curves, and produces a more magnified but dimmer image than that of a shorter-focal-length lens. The basic image from a lens is upside down.*

Choosing your supplier

Where should you buy your telescope? Choosing your supplier can be almost as important as choosing the telescope itself. There are several possibilities, listed here in order of increased preference:

1. From an ad in a newspaper or magazine offering "Special Purchase" or "Reader Offer."
2. From a catalog that sells everything from frying pans to frocks.
3. From a local camera shop.
4. From a specialist telescope company that advertises in astronomical magazines or on a website, or has a showroom with a wide range of instruments in stock
5. Direct from the manufacturer.

The first two options are very chancy. Even trusted newspapers with an international reputation for intrepid investigative journalism have buyers that don't know one end of a telescope from another, and certainly wouldn't know a bad one; it is the same with the catalog buyers. And some formerly trusted names in optics have been bought out by primarily marketing companies that will put the badge on any old rubbish for as long as they can get away with it, so you can't trust a name (or apparent country of origin) in a catalog.

Local camera stores often have a few telescopes available, and may have knowledgeable staff, but judge for yourself. At least it may be easier to try in advance, or return the goods, if you are not happy.

Most specialist telescope companies know what they are selling, but some are basically box-shifters and do not know too much about the individual instruments, so a quick phone call may be a good idea to find out which is which. As with most retail outlets, those offering a quality service are likely to charge more than the box-shifters.

The major manufacturers, such as Meade and Celestron, do not sell directly, but rely on a chain of dealers. However, the smaller manufacturers, who generally make instruments to order, will often only sell directly.

What should you expect from your chosen supplier? It is a sad fact that manufacturers have been known to let substandard or misaligned instruments slip through, and while it should be robust enough to withstand shipping, there is always the risk that your expensive telescope will give poor results when it arrives. Page 79 tells you how to test your instrument, and if you feel that it isn't performing as it should you may need expert help.

The ideal supplier will test each instrument on an optical bench before selling it, will be prepared to realign it should it be faulty when you get it, and will be close enough to your home that it is not a problem to return it. Take this into account when buying, particularly if you live in another country from the supplier. The economics of shipping can make a large difference to the final cost.

For example, a telescope that costs $1000 in the US may well be advertised by a supplier in the UK for nearly £1000, even when the pound is equivalent to nearly $2.00. These days it is a simple matter to order on your credit card and be billed in the direct equivalent in your own currency. But when the big box arrives at your door you will get another bill for duty and Value Added Tax. Even with these and the shipping costs, you will still have paid less than the UK retailer charges.

But if you find that the telescope underperforms, or, worse still, has a broken component, your only recourse may be to return it to the supplier at your own expense – with luck making sure that you don't have to pay the duty and VAT again because the supplier has neglected to tell the shippers that these are now your own goods. At this point your saving disappears, or at least has not been worth the hassle. Direct importing is safer for less fragile items such as eyepieces and filters, however.

Buying by credit card is always advisable, as you do get some consumer protection should the goods turn out to be faulty.

whole range of objects that would be invisible, or scarcely visible, using a 70 mm under the same conditions. Most of us are faced with less-than-ideal conditions, particularly where light pollution is concerned, and the extra aperture does help to bring in more objects and show more structure in them. For general hunting of faint objects, an 80 mm is a good compromise between having a telescope small enough to carry around easily, and still show something worthwhile.

The range of refractor sizes has changed considerably in the 21st century. Until about the 1980s, it was 3-inch (76 mm), 4-inch (102 mm), maybe $4\frac{1}{2}$-inch (114 mm) and 6-inch (150 mm). With the rise of manufacturing in the Far East, the range now jumps in 10 mm aperture bands, with some of the old sizes as well. And this is without considering the multiplicity of types of mounting that are currently available. To confuse the choice yet further, the market divides up into the traditional basic achromatic long-focus refractors, around f/12, and the more recent shorter-focus models, usually around f/5. What's the advantage of choosing a longer telescope, when a shorter one is obviously much more convenient?

As described on page 29, a telescope's focal length or focal ratio is as important a guide to its performance as its aperture. Bear in mind that the focal ratio is a guide to the stubbiness or length of a telescope, size for size. In a nutshell, you get wider-field views with short-focal-ratio telescopes, and more magnification out of the longer ones. This is true of all telescopes, and not just refractors. There is a limit to the magnification that you can coax out of any telescope, and short-focus instruments have their limitations. The shorter the focal ratio of a lens, the steeper the curves on its surface as the light must be refracted through a greater angle than with a longer-focus lens. This increases the false color, a problem that has traditionally affected refractors. The increased curvature makes the lenses behave like prisms, splitting light into its component colors.

▼ The 60 mm refractor in the foreground has a basic altazimuth mount, while the 80 mm refractor behind it has a more sturdy equatorial mount.

False color, or *chromatic aberration*, generally shows up as a blue haze around objects, particularly bright

◀ *The Sky-Watcher 80 mm f/5 refractor is compact and gives good medium-power views. A similar model is also available from Orion in the US.*

ones. Although refracting telescopes other than toys are called achromatic, meaning free from color, the term is only relative. They still suffer from some false color, unless you spend a considerable sum. So short-focus refractors are best suited to comparatively low-power views. Increase the magnification to study the planets and the results will probably be disappointing, both in terms of optical performance and false color.

Refractor devotees simply get used to the blue haze around planets, and point instead to the better contrast and legendary clarity of refractor performance (see "Lens vs mirror," page 41). Many users overcome the false-color problem by using the "Fringe-Killer" eyepiece filter made by Baader, which is claimed to improve the performance of any traditional refractor by selectively filtering out the blue wavelengths.

The Go To revolution

With the introduction of computer-controlled small telescopes, electronics is entering the cheap end of the telescope market. You can now buy small refractors that will, in theory, direct you to find a wide range of objects in the sky without any prior astronomical knowledge. The more advanced models have their own motors and will go to any chosen object automatically, so these are known as "Go To" models. Having found the object, a Go To instrument should then continue to track it through the sky, allowing you to view continuously without the need to push the telescope to follow it. The systems fitted to the smaller instruments, however, have no motors and require you to push them until the display indicates that you are on-target, so they are sometimes known as "push-to." From then on, you must follow the object through the sky manually.

Although this electronic control pushes the cost up, the innovation does have a strong marketing appeal. The biggest hurdle for most beginners has always been learning the sky and finding something worthwhile to observe. Now this barrier has been overcome – or has it? You often do not need any astronomical knowledge to set the

How to avoid a lemon

Many of the cheapest telescopes on sale in department stores, toy shops and other non-specialist outlets are non-achromatic. This means that they have a simple objective lens, giving images that suffer from bad false color. In an attempt to reduce this from diabolical to merely atrocious, the telescope has what is called an aperture stop inside the tube – a mask that cuts down the effective lens diameter. This also dims the image.

In the worst examples, the view through such a telescope is so bad that it shows virtually nothing more than can be seen with the naked eye. Saturn, for example, is just a blur. An easily seen double star such as Albireo, which should appear as two stars, is similarly indistinct. But no matter how well it is made, such an instrument can never show more than a very limited range of objects. It would be better to buy cheap binoculars.

People often buy these telescopes for their children, particularly when the child first becomes interested in astronomy, with the hope that if the interest grows it will be worth spending more later on. But all too often the child is so disappointed by the dim images that the spark is never kindled. So how can you spot the lemons?

Quite often, the specification is misleading, quoting a lens diameter of, say, 50 mm. However, although the piece of glass may well be 50 mm across, the stop behind the lens cuts the effective aperture down to maybe 20 mm. To see if this is the case, the trick is to look down the tube through the objective. Any stop will be only too visible.

Bear in mind, though, that good telescopes may have baffles, which are wide rings on the inside of the tube whose job is to prevent reflections from the tube wall from reaching the eyepiece. However, these will be much wider than an aperture stop. To be sure, look at the little circle of light at the far end of the tube. You should be able to see all of this when looking through the extreme edges of the objective.

Why don't trading standards authorities do something about the misleading description of telescopes? The main problem is a lack of knowledge on their behalf. They measure the diameter of the lens and find that it is indeed as specified, so they do nothing. But the point is that as a telescope objective, it's the clear aperture that matters. It is like describing a TV as having a 50 cm screen if in fact you only get a picture on the central 25 cm, and all the rest is blank glass. No one would get away with that – you are being misled by the implication that the picture, and not just the screen, is 50 cm across.

On the occasions when I have complained and persisted, I have been successful in making the telescope importer change their claims. In some cases, they were completely unaware that they were lying. But it's an uphill struggle to make sure that all such telescopes are correctly advertised. I urge anyone who is misled by package claims to complain and get their money back.

◀ *On the left, a good-quality 50 mm refractor with the full 50 mm aperture. On the right, a telescope claiming to be 50 mm but in fact stopped down to only 20 mm.*

Eyepieces and magnification

The magnification of a telescope depends on the focal length of the eyepiece in use at the time. Divide the focal length of the objective or mirror by that of the eyepiece to find the magnification. A 10 mm eyepiece, for example, gives a magnification or "power" of 70 if used with a telescope of 700 mm focal length (whatever type of telescope it is), or 200 if used on a telescope of 2000 mm focal length.

It is best to have a range of powers, usually at least three. To help achieve this, a Barlow lens is useful. This is a lens that multiplies the power of any eyepiece it is used with. A 2X Barlow, for example, doubles the magnification of any eyepiece it is used with. For more info on eyepieces and similar accessories, see Chapter 7.

telescopes up – all you need to do on Celestron models with SkyAlign, for example, is to point the telescope at three bright stars or even planets. You don't even need to know which stars they are.

Once the telescope knows the date, its location and the positions of its reference stars, it should then be able to find any other object in its extensive database for you accurately. However, the database probably contains objects that are completely invisible with such a small refractor under typical suburban conditions. Unless you know what to look for anyway, you may end up staring at a blank bit of sky unsure whether you have done something wrong, or the telescope is malfunctioning, or the object you are looking for is just too faint or too small to be visible in your light-polluted sky – or could never be seen anyway.

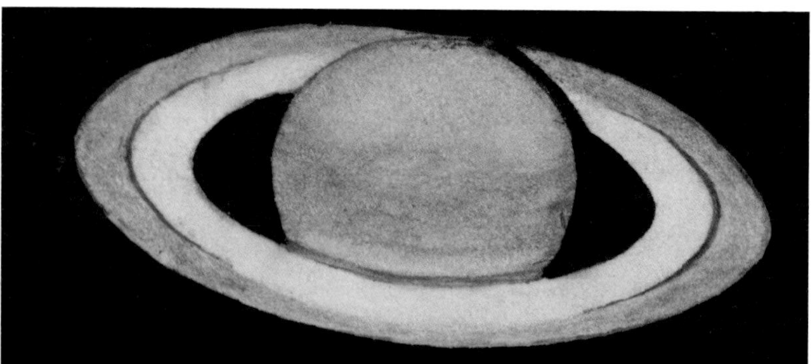

▲ *Saturn as seen with a 76 mm refractor, with a magnification of 80. The dark band in the rings is the Cassini Division, and the shadow of the globe on the rings is also visible. This drawing was made by Martin Lewis as a 14-year-old schoolboy in 1975, and shows just how much can be seen with a small aperture with good eyesight.*

When Go To works well it is a joy to use, and can really enhance your viewing pleasure, but it is not an instant solution to lack of astronomical knowledge. Some experienced observers are unhappy with the whole concept of Go To, particularly for beginners. This discussion relates not just to small refractors but to a wider range of instruments, so for more on the topic, see the website www.stargazing.org.uk.

▲ *The Meade ETX-80 is an 80 mm short-focus f/5 refractor with altazimuth Go To mount and tripod.*

While the more advanced amateurs may scoff at such small instruments, and point out that for the same price you can get a much larger manual telescope that will give much better views and show more objects, for many people there is a great appeal in having such a clever telescope at a reasonable price. The small computer-controlled refractor has brought telescopes to the masses.

High-performance refractors

A new wave of high-quality refractors has swept the market in recent years. Instead of being plain old achromats, with some residual false color, these are variously known as apochromatic, APOs or ED telescopes. They have significantly less false color than ordinary achromats, or even virtually none at all.

The reason for this is the choice of glass from which the lenses are made. To correct for false color, objective lenses are made from at least two separate components, or elements. Each element is made from a glass with a different light-bending power, or refractive index. Traditional materials allow only a certain amount of correction, becoming increasingly unsatisfactory at shorter focal ratios of f/7 or f/5. However, by using more exotic materials it's possible to make either long-focal-ratio lenses with improved color correction, or short-focal-ratio lenses. Such glasses are now becoming widely used for premium-quality lenses.

The 1990s saw the rise of the fluorite refractor. One element of the lens is made from a crystal of fluorite, carefully grown under controlled conditions, rather than glass. Fluorite refractors have legendary performance, with much reduced or negligible false color right across the spectrum, and costs to match.

How much can you see?

People often ask, when you show them your new telescope, "How far can it see?" This is a bit like asking the owner of a new car how far it can travel. Even a small telescope can see objects millions of light years away, but this is not a very meaningful figure and it is hard to be precise. Even professional astronomers can't be sure how far such distant objects are!

There are two important properties of a telescope – its light grasp and its resolving power. Both depend on the telescope's aperture – the clear diameter of the mirror or lens that gathers the light. The light grasp depends on its area, so it increases with the square of the diameter. A 50 mm diameter lens has a collecting area of about 2000 sq mm while a 100 mm diameter lens has nearly 8000 – four times as much.

Light grasp can be measured by the faintest star a telescope will show to the eye – what is called the limiting magnitude. This depends very much on the sky conditions and the observer's eyesight. It also refers only to stars, whereas most people are more interested in distant galaxies and faint nebulae, whose light is spread out over an area of sky. It is quite possible to see close to the limiting magnitude of a telescope, even from an indifferent suburban observing site, yet be unable to see a fairly easy extended object on the same occasion. The contrast between the faint object and the sky background is what counts here, as it does when observing detail on planets, yet there is no easy way to measure the contrast of an image.

The other property, resolving power, is a measure of the finest detail your telescope will show. Traditionally this is measured by finding the closest double star that the telescope will just show as two separate stars. There are many double stars in the sky, where two stars are in close orbit around each other.

The table opposite shows the limiting magnitude and resolving power of various telescope sizes, but this doesn't mean much out of context. To give you a feel for it, consider some typical objects that you might want to observe. There is little interest in seeing faint stars as such, but Pluto, the distant dwarf planet, is a starlike point of around 14th magnitude. The brightest quasar (a type of starlike distant galaxy) is about 13th magnitude. The faint stars in globular and open clusters are usually fainter than 12th magnitude, so the larger your telescope the more interesting they will appear.

Many of the brighter galaxies in the Virgo cluster are around 9th or 10th magnitude, so they should be within the reach of any telescope on this table, but you generally need a whole magnitude in hand where extended objects are concerned. Many run-of-the-mill comets are around magnitude 12 or 13, so you need a 200 mm telescope or thereabouts to see them at all.

When considering resolving power, bear in mind that the seeing plays a large part in what you can see, whatever your telescope. Professional astronomers measure seeing by the amount of blurring of a star image, while amateurs have more general scales of measurement. By looking at double stars of known separation, however, it is possible to estimate the seeing. Professionals aim for seeing better than 1 arc second, and consider half-arc-second seeing to be exceptional, even for a top-quality mountaintop site. From an average site, a figure of 2 arc seconds or worse is typical. Yet this is the resolving power of a 60 mm telescope! Does this mean there is no point in getting a telescope larger than this for seeing fine detail?

Actually, when the seeing settles down, even from a city location it is possible to get sub-arc-second seeing. By using your eyes, and waiting for those moments of steadiness, you can see details with a modest telescope that are hard to photograph even with large instruments.

To give you an idea of how the planets would look with different telescopes, Mars, Mercury and Uranus are only 3 or 4 arc seconds across when they are at their most distant. But Mars when closest is 25 arc seconds across, while Jupiter and Venus are more than twice that size. Half an arc second on the Moon is about 1 km, so if anyone asks you if you can see the tracks of the Apollo Lunar Rover on the Moon, look at the table and you will realize that even a racecourse on the Moon would be only a tiny speck with a large amateur telescope.

Aperture (mm)	(in)	Limiting magnitude	Resolving power (arc seconds)
50	2	11.2	2.3
60	2.4	11.6	1.9
80	3.1	12.2	1.4
100	4	12.7	1.2
115	4.5	13.0	1.0
150	6	13.6	0.8
200	8	14.2	0.6
250	10	14.7	0.5
300	12	15.1	0.4
400	16	15.7	0.3

Other materials, such as the ED (Extra-low Dispersion) glass from which the majority of such lenses are now made, are less expensive. The ads often refer to an "ED Achro" or "ED Apo," for example. An achromatic lens is one in which false color is fully corrected at two points in the spectrum, while an apochromatic lens is corrected at three points, and should be effectively color-free.

Initially a very expensive item, the ED refractor has come down in price over the years. You can get a variety of Far Eastern ED refractors, marketed under a variety of names, which perform very well indeed against top-of-the-range instruments. While they do not have quite the same color correction as the more expensive ED refractors of the same size, they still have excellent performance and are much cheaper. Even so, the telescope alone (known as the optical tube assembly or OTA) costs as much as a traditional 120 mm refractor on an equatorial mount.

▲ *The Sky-Watcher 80 mm ED refractor, shown here complete with eyepiece, diagonal, finder and dovetail, with Crayford focuser. Similar instruments are available as tube (OTA) only from outlets such as Orion Telescopes and Binoculars (US).*

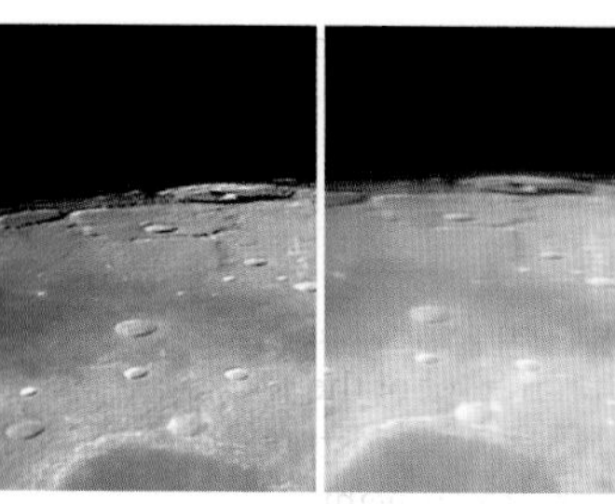
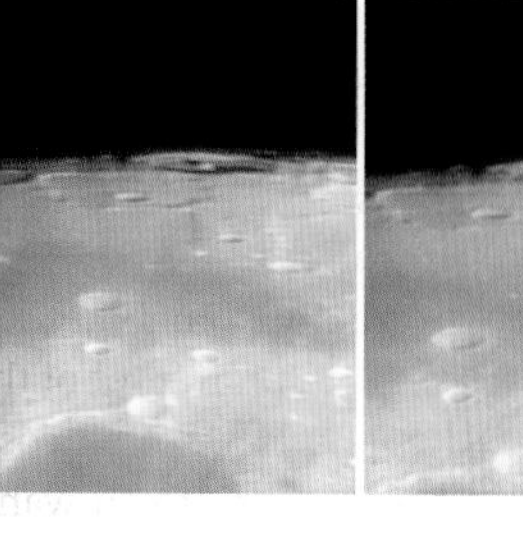

► The lunar crater Pythagoras photographed using three small refractors. Left to right: Sky-Watcher 80 mm ED f/7.5, focal length 600 mm; Unbranded 80 mm f/7.5, f = 600 mm; Bresser Skylux 70 mm f/10, f = 700 mm. The ED instrument has no false color at this scale, while the cheap refractor with the same focal length has strong color and shows much less detail. The f/10 Bresser, with longer focal length, also has strong false color but shows some detail because of its longer f-ratio.

When buying an OTA alone, incidentally, check before buying whether it requires a star diagonal. The focusing range of many refractors is not great enough to allow you to focus both with and without a diagonal. This means that every view you get is laterally inverted, like it or not. The information is rarely mentioned in ads, nor even in reviews.

Typically, an ED or APO refractor will have a Crayford-style eyepiece focuser, usually capable of taking 2-inch eyepieces (see page 149). This offers finer focusing control than the standard rack-and-pinion focuser, and often has two focusing knobs, one for coarse and one for fine focusing. There is also often a locking screw, which is invaluable when you are photographing through the telescope using a heavy camera and are pointing it at an extreme angle. The weight of the camera can easily pull the focuser out from where you have spent ages getting the best focus.

A refractor for photography

As well as advances in optical design, the advent of digital photography and availability of new filters has meant that astrophotography from the suburbs is once more possible. Amateur astronomers are routinely taking photographs from urban areas which once would have required dark skies and supreme effort. Many of these photographs are taken not through large telescopes but using small, short-focal-ratio refractors, in the size range 60 to 90 mm. These are well suited to wide-field photography of nebulae, star clusters and the nearer galaxies, for example. In particular, they have a wider good field of view than a reflector of the same focal length.

However, there are limits. When the focal ratio becomes shorter than about f/7, what's called field curvature can pose a problem. The image plane of a telescope is rarely completely flat, but is curved. For purely visual observing a small amount of field curvature is not a problem, but the image plane of cameras is flat and any field curvature shows up as defocused star images at the edges of the field of view when the center is sharp. At the advanced end of the telescope market there are refractors especially designed to give flat fields for wide-field photography, sometimes known as astrographs, and you can also buy devices that fit in front of the camera to flatten the field of lesser instruments, but these often cost a considerable fraction of the cost of the telescope itself. If you want to buy a small refractor for photography, either take advice from a specialist supplier or take a look at online galleries of photographs and see what others have used.

Refractor limits

Not so long ago, the larger refractor market stopped at about 120 mm except for a few highly expensive and legendary fluorite instruments up to 175 mm. But 150 mm refractors in both long and short focal lengths are now available under a variety of names, and very competitively priced.

Fluorite or ED refractors are made in sizes up to at least 200 mm, with price tags, when you include a decent mounting, similar to those of a family car.

At the larger refractor sizes, one starts to question the advantage of a refractor over a reflector or catadioptric of similar or somewhat larger aperture, which will perform as well for much of the time but at significantly lower cost. Refractor devotees point to their image brightness, excellent contrast, lack of required maintenance, ability to give good images with little cool-down time, and excellent wide pin-sharp flat fields of view that are particularly suited to deep-sky astrophotography. But there is a price to pay for this, and owners of top-of-the-range refractors from manufacturers such as Takahashi or Astro-Physics usually know what they want from a telescope and are happy to pay for it.

► *Galaxy M31, photographed by Ian King through a 75 mm Pentax telescope using a Starlight Xpress CCD camera.*

Much has been said about the relative merits of reflectors and refractors, but as aperture increases, reflectors tend to win the argument. However, a good refractor with a well-corrected lens should give excellent performance, unrivaled by a reflector of the same aperture.

Reflecting telescopes

The second half of the 20th century saw the rise of the large reflecting telescope as a common amateur instrument. Today, amateurs regularly use telescopes that would not have been out of place in a professional observatory before the 1970s. But small reflectors have improved as well. At one time, a reflector smaller than 150 mm was regarded as pretty useless, whereas nowadays many amateurs begin with such an instrument.

There is a practical limit to the smallest reflecting telescope, which is set by the need to reflect light to the side of the tube using a small flat mirror, known as the secondary. Scale down a large telescope, and the beam of light reaching your eye will be minute, giving a very dim

▼ Two Sky-Watcher 150 mm refractors compared. At left, the f/8 model, with a focal length of 1200 mm, and at right the f/5, 750 mm focal length version. Both are on identical HEQ-5 equatorial mounts. The f/5 instrument gives excellent low-power wide-field views, and will take 2-inch eyepieces. The f/8 model is better for higher-power work, particularly when used with a minus-violet eyepiece filter.

Lens vs mirror

The debate over whether refractors are better than reflectors and vice versa enlivens many a long winter evening. Fashions and opinions change, and some old dogmas have been exposed.

It was once believed that quite a small refractor was the equal of a considerably larger reflector, in performance on detail if not in brightness of image. That myth has pretty well been laid to rest, but people now claim that their neat little portable Maksutov outperforms much larger telescopes. So what's going on?

Because refractors have no central obstruction, the contrast of their images will always be better than any reflector, whether it be Newtonian or catadioptric, with the rare exceptions of some specially made unobstructed reflectors. We are assuming here that the instruments are well made and perform as they should. The brightness of a reflector's image is also reduced by the central obstruction and by the fact that the reflectivity of the coatings is only about 88 per cent per reflection, while the transmission of a multicoated objective lens is around 90 per cent in the smaller sizes.

But the difference in performance is by no means as marked as was once believed. As a rule of thumb, subtract the obstruction caused by the secondary from a reflector's aperture and you have roughly the equivalent size of refractor. So a 150 mm reflector with a 35 mm obstruction is about as good as a 115 mm refractor.

Also in favor of refractors is their ruggedness, with no mirrors to tarnish or get out of line. They are less likely to be troubled by air turbulence within the tube, as it is sealed, and the light travels through it only once rather than twice or even three times as in reflectors and catadioptrics. Astrophotographers approve of high-quality apochromatic refractors for their wider fields of view without distortion.

But the major drawbacks of refractors are their false color, which persists to a certain extent even in some models made using special glass, and their expense, aperture for aperture. With their lenses at the top of the tube, they are prone to dewing up.

Reflectors, on the other hand, are free from false color, are cheaper for a given aperture than refractors, and can be made in sizes and shorter focal ratios that would become prohibitively expensive in refractors. But they need more maintenance, though catadioptrics with their sealed tubes should suffer less from tarnishing of the surfaces than Newtonians. A Newtonian with a standard tube (that is, not an open or skeleton tube) will be much less likely to dew up than a refractor or catadioptric, however.

One major reason for the claimed superior performance of refractors, and these days Maksutovs, lies not in their optical quality but simply in their smaller size. Atmospheric seeing acts a little like frosted glass windows – the sort with fluting or blobs, rather than a fine-ground surface. From a distance, when you are looking through a large area of the window, you have a very poor view of what lies beyond. However, get close-up and peer through a small section, and you can actually see through it rather better.

Seeing is caused by air cells of different densities, constantly on the move. These cells are of the order of a few centimeters across, near ground level, so a small telescope, whether refractor, reflector or Maksutov, will see through individual cells. But a larger instrument, of whatever sort, often looks through several cells at once, resulting in blurring or even multiple images. When the seeing settles down, however, the larger telescope really comes into its own and reveals more detail. Generally speaking, larger telescopes are reflectors rather than refractors, hence the reports that a small refractor or Maksutov can outperform a larger reflector or SCT.

How a reflector works

There are two main ways of changing the path of light – by refraction and by reflection. Even a simple shaving or compact mirror with a concave surface will focus light, as you can prove for yourself by reflecting the light from a window on to a piece of card. The advantage of using a mirror rather than a lens is that all colors of light are reflected equally, so the image has no false color.

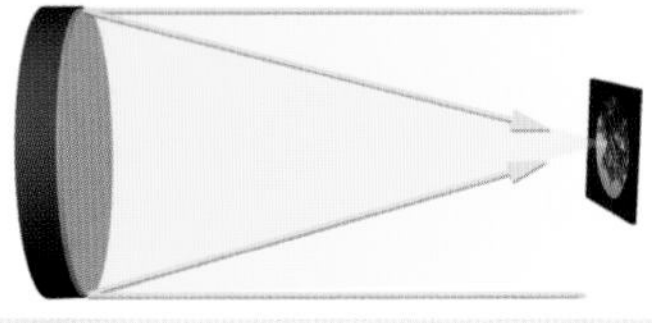

But one drawback is that the image is reflected back into the path of the incoming beam. So in a practical telescope, the image then has to be reflected to a point outside the tube where it can be studied with an eyepiece, as in a refractor. The normal way of doing this is to use a small flat secondary mirror to reflect the image to the side of the tube – the classic Newtonian system, first designed by Isaac Newton in 1668.

Almost all reflecting telescopes, therefore, have secondary mirrors in their incoming light beams, which degrade the image slightly compared with the unobstructed beam of a refractor. The phenomenon of diffraction – the slight bending of light around an obstacle in its path – means that the contrast of images is reduced by a central obstruction. The larger the obstruction, the worse the effect. As long as the obstruction is less than 20 per cent of the aperture, however, the loss of contrast is hardly noticeable. The secondary mirror also cuts out some of the light, so a reflector gives a slightly dimmer image than a refractor of the same aperture, particularly as the mirror coatings tarnish over time.

The spikes (usually four) on star images in photographs are caused by diffraction around the vanes of the "spider" that holds the secondary in place in a Newtonian.

The accuracy of manufacture of a mirror is often described as "diffraction limited" or "quarter wave." This implies that the performance of the telescope is limited only by the effects of diffraction, not by poor manufacture. However, this in itself is not a well-defined term. The phrase

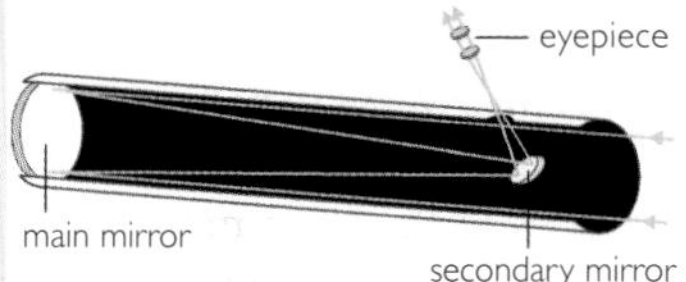

"quarter wave" should refer to the accuracy with which light is brought to a focus – within a quarter wavelength of the ideal, the wavelength being that of yellow light. This means the mirror should be manufactured to within one eighth-wavelength of the ideal figure, since reflection doubles the error.

Practical tests suggest that while many regard quarter-wave optics as diffraction limited, it is just possible to tell the difference between quarter- and tenth-wave optics under good seeing conditions. On the other hand, there is no point in paying for even more perfect optics than quarter-wave unless you plan to observe planets regularly and expect the observing conditions to be excellent.

Notice that a mirror made to within a quarter wavelength of the ideal figure will actually have a half-wave error at the focus point, which is unacceptable. So read the claim carefully, and make sure that the accuracy refers to the wavefront error at the focus, rather than the error in the mirror's surface.

The concave surface of the main mirror should be parabolic in section, but small long-focus mirrors may be spherical, which is easier to make, with no noticeable loss of quality. Small short-focus mirrors, however, should be parabolic for acceptable results.

image. The smallest practical reflecting telescopes are 76 mm aperture, and simple Newtonian reflectors of this size are often found in camera shops and catalogs, sometimes adorned with impressive names. While they may have acceptable optical performance, they suffer from the problem of all such telescopes made to a price of being rather lightly mounted and tricky to use, as with the cheapest small refractors. It's when you come to use them that you realize how useful "slow motions" are – knobs that allow you to adjust the pointing of the telescope. Instead there may be a screw adjuster on a steadying arm, with a small range of travel only. This is better than nothing, but if it is stiff to turn it is worse than useless. Just finding your target and keeping it in the field of view calls for great patience and indeed skill. Again, like their small refractor cousins, some telescopes are equipped with impossibly high magnifications.

If you stick to the lower magnifications, or have the patience to keep the scope steady if you view at higher power, these small telescopes can give you rewarding views of the Moon and many deep-sky objects. One tip is to put a heavy weight on the accessory tray between the tripod legs, which may help to steady the instrument a little.

The market really begins at around 90 or 100 mm aperture, and at this size there is a wide variety available, from the cheap and cheerful to the precise and pricey. Most of the major manufacturers have starter instruments of 114 mm – that is $4\frac{1}{2}$ inches – and this is an excellent bridge between the small refractors and anything more advanced. In my view a well-made reflector of this size can rival a 75 mm refractor costing much more. It is a pity that the demand these days is for compact instruments. With reflectors as with refractors, the longer the focal ratio, the easier the optics are to manufacture and the better image you should get for your money.

A telescope with an aperture of 114 mm, while by no means large, begins to open a real window on the Universe. You can see meaningful

▶ *The Sky-Watcher Explorer 130 is a 130 mm f/6.9 Newtonian reflector. It will show a wide range of objects, and its equatorial mount makes it easier to track objects as they move across the sky.*

detail on the planets, given good conditions, and can cast your net wide in search of deep-sky objects. No longer are you restricted to the nearest galaxies and brightest nebulae. This is the minimum size of telescope that a serious amateur astronomer can use, and these telescopes are by no means expensive, at around the price of a small TV set.

Only a few years ago, the standard 114 mm Newtonian reflector was a basic f/8, but now the manufacturers have shortened their focal ratios, presumably to make their telescopes shorter and more portable. There are two types, one with a focal ratio of f/5, while the other is f/8.77. Yet both have equally short tubes. What's going on?

Both telescopes have mirrors of short focal length, but the one that claims a focal ratio of f/8.77 has an additional lens in the eyepiece barrel – essentially a Barlow lens – which increases its working focal length to 1000 mm. So you can get high magnifications, but unfortunately you don't have the choice of also getting wide-field views from the essentially short-focus mirror. The addition of the lens allows them to be described as catadioptric, but these are not in the same league as Maksutovs or Schmidt-Cassegrains, which have genuine corrector plates. One such model I tested gave rather poor images at high power because the additional lens was of limited quality. It would be much better to buy a short-focus telescope in the first place and use a Barlow if you want higher magnifications – then at least you have the choice.

Regrettably, the standard classic long-focus 114 mm reflector is no longer widely available, apart from the venerable f/7.3 Russian-made TAL-1, which comes with a sturdy mount and pedestal which are much more stable than their Far Eastern counterparts in the same price range.

Occasionally you may come across short-focus small reflectors designed for wide-field low-power use only, maybe called either "comet catchers" or "Richest Field Telescopes" (RFTs). These are designed to give the widest possible field of view, and they are particularly suited to viewing or even searching for comets. Usually these are unsuitable for high magnifications, but they can give dramatic views of starfields in dark skies, which is their proper environment. At one

◄ *The f/14 Meade ETX-125 Maksutov-Cassegrain telescope provides portability with Go To, and its aperture is large enough to reveal a wide range of objects. It weighs 11.4 kg (25 lb) with tripod.*

time it was a fine amateur sport to search for new comets, but the advent of surveys for potential asteroid threats has reduced the chances of a discovery by purely visual means.

Small cats

There are other telescopes, of similar tube size, that are true catadioptric instruments. They have a corrector plate at the top end, which modifies the light beam reaching the main mirror, thus giving good-quality images from a short tube. The best-known examples are the Meade ETX series, which includes the ETX-90 and ETX-125, and the Celestron NexStar 4, all Maksutov-Cassegrains. Similar Chinese-made instruments are available under a variety of trade names. These telescopes have their eyepieces at the bottom. The focal ratio is typically between f/13 or f/15, which is fine for high-power work. These telescopes can give excellent planetary views, though they are not ideal for low-power wide-field views such as you get from RFTs.

The Maksutov design is capable of high optical performance, so you should expect to see very good-quality images from these telescopes. They are particularly suited to observing the planets and some deep-sky objects, where you need good optical quality combined with compactness. Many observers prefer these neat and versatile instruments to refractors of similar aperture. But, as always, a good big 'un will always outperform a good little 'un. If you want to see real detail on the planets, you will probably not be happy with a telescope this small.

Larger reflecting telescopes

For years, the standard amateur telescope was the 150 mm or 6-inch reflector, and even now, with the ready availability of SCTs of larger size, it is still an excellent instrument. You can't go far wrong with a 6-inch reflector, though these days it is regarded as rather small. But if you want a dependable, versatile telescope that will show you most things you demand of it, yet is inexpensive and not too bulky, a 150 mm reflector is a good choice.

It is probably not going to be compact and

► *A basic 200 mm reflector such as the Orion Optics Europa 200 f/4.5, on an undriven Vixen GP-2 mount, has over twice the light grasp of the ETX-125 at a lower price but without Go To.*

◀ *Russian TALs have many fans for their optical quality and engineering, though they are heavier than Far Eastern instruments. Shown here is the 150 mm f/5 TAL-150, with solar projection screen.*

portable, however, and will not be the sort of instrument you can take with you as airline baggage. As ever, the longer the focal length the better chance you have of getting a good mirror at a reasonable price. An f/8 mirror gives a tube length of 1200 mm, which is not excessive and brings the eyepiece (at the top of the tube, remember) to a convenient height above the ground for observing. An f/6.3 tube will be about a meter long. By the time the mirror gets to f/4, the tube is distinctly stubby, but such an instrument is likely to have a tendency to drift out of collimation – that is, the mirrors become misaligned, resulting in distorted images – unless the optics are very well secured.

Reflectors are by no means perfect telescopes. As you move away from the optical axis (the center of the field), the aberration known as coma becomes increasingly noticeable, with stars appearing like little comets. For most practical purposes it is acceptable in the longer f-ratios, and can be reduced by a device known as a coma corrector.

Commercial reflectors are available in sizes up to 250 mm from the major manufacturers, and the sky's the limit from the smaller companies, with 400 mm being typical. At the larger sizes, the mounting is an important consideration as it has to provide accurate driving and freedom from vibration. Equatorially mounted instruments larger than 300 mm are really best suited to an observatory. The cost is also considerable, even if Newtonian reflectors are cheaper, size for size, than SCTs. And this is where the Dobsonian comes in.

Dobsonians

A Dobsonian mount is a very basic method of putting a telescope together on a simple altazimuth mounting. In its original form it is heavy, which is no bad thing where a telescope is concerned. A good Dobsonian is free from those fascinating but irritating dance routines

How a catadioptric works

Catadioptric telescopes combine reflecting and refracting components. The basic image is produced by a steeply curved mirror, which would normally suffer from poor image quality at the edge of the field of view, but this is compensated by a corrector plate at the top of the tube. The curvature of the surface is only slight, so any false color introduced is negligible.

The most common catadioptric is the Schmidt-Cassegrain telescope (SCT), which uses a secondary mirror to reflect light back down the tube through a hole in the center of the main mirror. This gives a comparatively long focal ratio. The secondary mirrors of SCTs are fairly large, resulting in poorer optical performance than for an equivalent size of Newtonian reflector.

The other common design is the Maksutov, which has a more steeply curved corrector plate than an SCT. The optical performance is very good, but over a restricted field of view. The steeply curved corrector is tricky to make in larger apertures.

Both SCTs and Maksutov telescopes have viewing positions like those of refractors, in line with the tube. They have much more compact tubes than for the equivalent size of reflector or refractor.

that a star can display in some lightweight telescopes. With those, touch the eyepiece with an eyelid and the star is off on an enthusiastic vibration, performing exquisite circles and ellipses of ever-decreasing size until it settles down again. In the slightest breeze, the image is never still. But in a Dob, you can often happily refocus without so much as a tremble of the image. The wooden construction damps out vibration.

The downside of the Dob is that you are constantly shifting the tube in order to track an object. Making a detailed drawing of a planet can become a trial of patience. And locating objects near the zenith can be very frustrating, because you may need to swivel the telescope through 180° simply to move it a few degrees across the sky.

Dobsonians come into their own in sizes larger than 150 mm, but at this size they remain

► Dobsonians provide stability at low cost and are the cheapest way to observe a wide range of objects. This is the Sky-Watcher Skyliner 250PX, the 250 mm version. FlexTube models can be shortened for easy transport. The range also includes 150, 200 and 300 mm instruments, plus digital setting circle options.

quite easy to manage and indeed to make for yourself. If you really want to save money, buy just the optics of a telescope and build a Dobsonian mount. As long as you are half competent with a saw, a drill and a screwdriver, you should be able to run up an excellent telescope in a few hours using readily available materials.

Dobsonians are commercially available in sizes up to at least 450 mm, though not necessarily off the shelf. Telescopes this size are still transportable, because the instrument can be disassembled so that it will fit into a car. For many deep-sky observers, their Dob sees light only on expeditions to dark-sky sites where they avidly track down faint galaxies. Though the Dob is essentially a low-cost instrument for its aperture, there are also precision-built versions that may cost as much as an SCT of the same aperture, offering great compactness when disassembled and top-quality optics.

One problem with a Dobsonian is that you are pretty reliant on your own resources when it comes to finding the objects in the sky, and a good finder is essential. But the inevitable has happened, and it is now possible to get Dobsonians with optical encoders that provide a "push-to" facility, with readouts that tell you exactly where in the sky you are pointing – also known as digital setting circles. Sky-Watcher now also have a motorized and auto-tracking Dobsonian mount with collapsible tube that will fit easily into most cars, further improving the design.

SCTs – king cats

At one time, the good old Newtonian reflector reigned supreme. But now, the title of "King of the Telescopes" undoubtedly goes to the Schmidt-Cassegrain telescope, or SCT, which is a catadioptric design. These instruments first came on the market during the 1970s, became market leaders among medium-sized amateur telescopes in the 1980s, and developed into highly versatile instruments in the 1990s. Since about 2007, Meade have altered the design of their Schmidt-Cassegrains to a version known as ACF, meaning Advanced Coma Free. It is still basically an SCT design, but claims smaller distortion due to coma near the edge of the field of view, and is treated here as a Schmidt-Cassegrain design.

Today, the 200 mm or 8-inch SCT is the standard telescope. It offers a reasonably large aperture coupled with a tube of manageable size, all in a neat package that includes a driven equatorial mounting with the potential for photography and CCD imaging. All recent versions have their own built-in computer control that will find faint objects with the minimum of setting-up. It's also possible to link them to a home computer so that they can be directed by a click of the mouse, and even programmed to search galaxy after galaxy in an automatic search for supernovae.

So is the SCT everyone's ideal telescope?

Let's take a closer look at the SCT's good and bad points. In its favor, it can give color-free images, and has a compact tube which is sealed, so little maintenance of the mirror coatings should be needed. Although the basic 200 mm aperture instrument is f/10, which can mean a fairly narrow field of view, focal reducers are available that will shorten the effective focal ratio to f/6.3, which gives a good wide field. This is enough to show most of the Pleiades cluster in one view, using a low-power eyepiece, rather than just a few stars.

The original SCTs were invariably on fork mountings, but now they are equally available on fork or German-type equatorial mountings. If the mounting is a fork, it can usually be used either as a basic altazimuth or converted to work as an equatorial, in which case you need what is called a "wedge." This allows you to tilt the mounting to match your latitude. The wedge is essential if you want to do long-exposure astrophotography. Recent models all include Go To motors and handsets, run from a clutch of AA cells or an AC adapter. Earlier versions, however, simply have a motor-driven base that is adequate for keeping an object in view. The whole assembly usually fits on an aluminum tripod.

The chief drawback of the Schmidt-Cassegrain design is that the optical system can never give as good results as a Newtonian of the same aperture, or a refractor of somewhat smaller aperture. This is because of the size of the secondary mirror, which introduces substantial diffraction effects. If you could compare the view of Jupiter, say, through an SCT with that through a Newtonian, you should see the same details but they would be noticeably less contrasty in the SCT – that is, the darker markings would be less dark. These drawbacks apply equally to Maksutovs, though Maks are renowned for their superior image quality across the field of view.

▶ Meade's 200 mm f/10 LX-200 is a best-selling SCT. The latest model comes with GPS position-finding as standard, to improve the positional accuracy of its Go To system. The LX-200 range includes apertures of 178, 203, 254, 305, 356 and 406 mm. This model is shown on the basic altazimuth fork mount.

When viewing a star even under perfect conditions, the dot of light shown by the SCTs is likely to be larger than that shown by a Newtonian, using the same magnification, because the optical system increases the size of the diffraction disk – a topic covered more fully in the box "How a reflector works" on page 42. However, these effects are not overwhelmingly bad in a well-made instrument, and are certainly not enough to put off the large numbers of discerning viewers who use SCTs nightly.

The design of the fork mounting means that when you are viewing an object in the general direction of the fork arms (overhead when used as an altazimuth, or around the pole when an equatorial), it's impossible to look directly through the eyepiece so you must use a star diagonal instead. This is something you get accustomed to, and many people use the diagonal all the time. The ads generally show the telescope with its diagonal in place. You lose a little light by using the diagonal, and also get a mirror image view, which can lead to problems if you forget it's happening. For the same reason, it can be impossible to take photographs of objects near the pole if the camera does not clear the mounting. These difficulties are overcome if the telescope is on a German equatorial mount, which is generally preferred for astrophotography.

The large exposed corrector plate of an SCT is very prone to collecting dew, and unlike refractors, SCTs do not have dew shields as standard. They are available as one of the scores of add-on goodies advertised in astronomy magazines. It may seem strange that a top-of-the-range telescope package costing as much as a car should not include a plastic dewcap whose real cost is little more than the price of a hamburger, even though the cheapest toy refractor has one, but this practice seems to be pretty universal among manufacturers.

Another common fault with SCTs is play in the focuser. To focus the telescope you turn a knob beside the eyepiece, and this actually moves the entire main mirror up and down the tube. Moving its whole weight by turning a single screw thread calls for careful engineering, and on many instruments play is common, with a considerable dead zone where nothing happens.

In order to provide a wide focusing range, the focusing mechanism may have a fairly coarse thread. Coupled with the play, this can make SCTs characteristically more difficult to focus than refractors or reflectors.

The internal optical parts of an SCT are fixed and are not designed to be adjusted by the user, unlike those of many reflectors. If they go out of alignment – or if they are misaligned when you receive the instrument, which is sadly not uncommon – realigning or collimating

◀ *The Meade 250 mm LX-200 is shown here with the fork mount sitting on the wedge, which allows the instrument to be used for long-exposure astrophotography.*

the instrument may need to be carried out by either a specialist or someone who knows what they are doing. This is a point to bear in mind when choosing your supplier. However, the alignment of the secondary mirror is adjustable after you have removed a cover, and most misalignments can be corrected in this way.

The 200 mm or 8-inch SCT is by far the most popular "serious" telescope in the world, with instruments also available in sizes up to 400 mm. Many instruments now include a GPS receiver which automatically sets the telescope's precise location and time. While it may seem that there is not much advantage in having GPS for locating a 355 mm telescope that can hardly be moved, the facility for obtaining the precise time improves the accuracy of the Go To operation.

Other Cassegrain telescopes

Although the SCT remains the most popular large telescope, Maksutovs are probably just as widespread these days because of the large numbers of Meade ETX telescopes. Maksutovs are now becoming popular in larger sizes, from 150 to 250 mm, and these are particularly prized for their good optical correction. They are usually around f/13 to f/15, so they are not suited to wide-field viewing.

Another variant is the Schmidt-Newtonian, which has a Schmidt corrector plate but a flat diagonal mirror instead of the secondary, so they have better coma correction than ordinary Newtonians but with a shorter tube, though the larger secondary assembly reduces the image contrast somewhat. Meade have 152, 203 and 254 mm models. The Schmidt-Newtonian design is particularly suited to deep-sky observing and photography.

Russian manufacturers are prominent in this area, with Intes-Micro a major player in the field. As well as the more familiar Maksutov-Cassegrains, there are also Maksutov-Newtonians, which have Maksutov corrector plates at the top end but Newtonian rather than Cassegrain secondaries. This combination produces focal ratios of f/5.5 or f/8 with very good optical correction and with quite small secondary mirrors for maximum contrast. The Sky-Watcher Explorer 190MN Pro is an example, at f/5.26.

Telescope mountings

The mounting of a telescope is as vital to its performance as its optics. Many telescopes come as packages that include a mounting, and many instruments, such as SCTs and Maks, are not readily detachable from the mounts with which they are sold. However, there is an increasing trend for selling instruments as the tube (OTA) only, and equally you can buy the equatorial mounts separately, such as those made by Synta and Vixen. These mounts are usually German-type equatorials, but Synta (Sky-Watcher) also have a range of the popular single-arm altazimuth Go To mounts with dovetails for small instruments. The Synta mounts are also used by other suppliers, such as Orion (USA), while Celestron, which is owned by Synta, have re-engineered the range so that they now have a separate range of mounts.

The Synta equatorials have designations starting with EQ, as in EQ5, and these designations have been "borrowed" by other Far Eastern manufacturers to describe their own clones. The Celestron mounts have designations starting CG or CGE. The largest Synta mount, the EQ6, is claimed to take loads up to 25 kg. Both Celestron and Synta have their own Go To systems.

The Japanese manufacturer Vixen transformed telescope mountings when they introduced the Super Polaris in the early 1980s. This had silky-smooth axes and stepper motors that gave fingertip control of the telescope position using a handbox with four buttons. It also included an accurate built-in polar alignment telescope with scales to allow for quick and easy polar alignment for astrophotography (see page 94). The Super Polaris was superseded by the Great Polaris (GP), which in its GP-2 update takes instruments up to 9 kg in weight, and there is also the GP-D2, which will take loads up to 11 kg. The Synta EQ5 is regarded as a clone of the Vixen GP mount. The Celestron CG-5, however, also exists in a Go To version.

In 2004 Vixen introduced a new mounting, the Sphinx, with a loading weight of 8.7 kg and a unique Go To system with a visual display handset which shows a color sky map – which in later versions is dimmable for night viewing.

For larger telescopes there are specialist mount manufacturers, such as Losmandy, whose largest mount will take loads up to 45 kg.

Attaching one manufacturer's instrument to another's mounting often requires an adapter plate. The smaller Synta mounts just take tube rings, but the larger ones, from the EQ5 upward, will take the

Telescope mountings

There are two basic types of mounting – the altazimuth and the equatorial. The altazimuth is the simpler and cheaper, and has two axes that allow you to move the telescope up and down (altitude) and from side to side (azimuth). This is fine for daytime viewing, but astronomical objects generally move at an angle to the horizon rather than just horizontally or vertically.

The equatorial mount also has two axes, but one (the polar axis) is tilted so that it is aligned with the Earth's axis. This means that the daily turning of the Earth, which causes the movement of most objects through the sky, can be counteracted by one motion only, that can be easily motorized.

Most refractors and Newtonian reflectors are mounted on a German-type equatorial, requiring a counterbalance that adds to its weight. It can be tricky to get the telescope in perfect balance, particularly when using heavy cameras or eye-pieces, but it allows access to all parts of the sky.

Most SCTs are on fork mounts, which need no counterbalance but have the drawback that access to the region around the pole of the sky is restricted, though fortunately this is not an area of major interest.

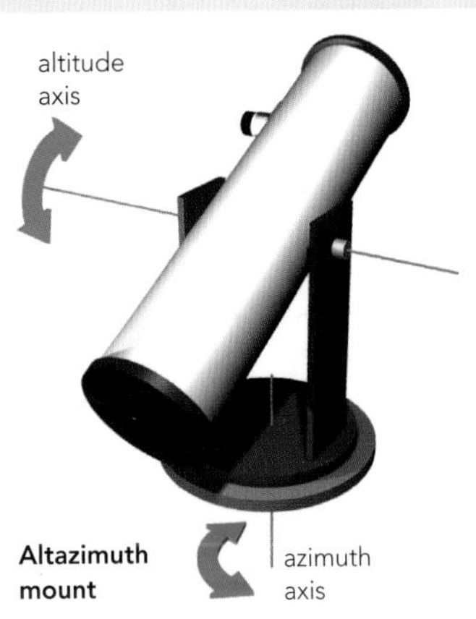

Altazimuth mount

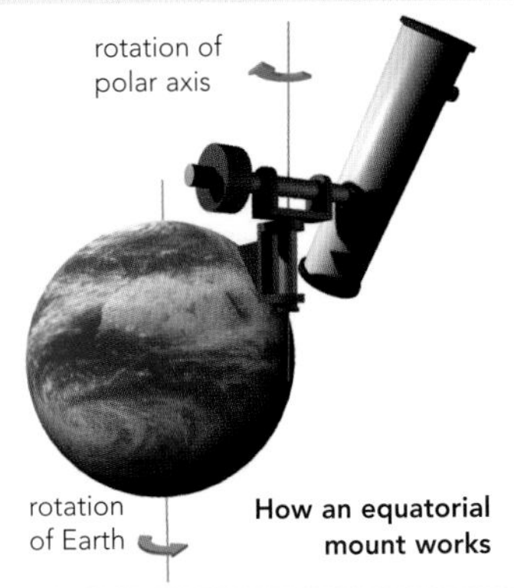

How an equatorial mount works

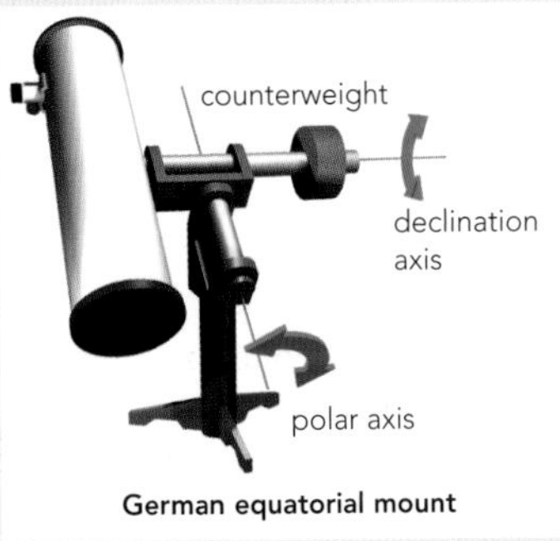

German equatorial mount

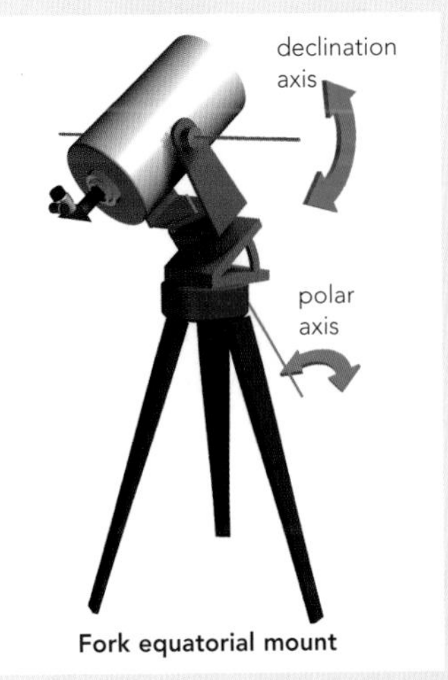

Fork equatorial mount

same dovetail plate that fits Vixen telescopes. This dovetail is pre-drilled with mounting holes for tube rings, and separate holes of the standard photographic tripod thread of ¼-inch, 20 threads per inch, known in the UK as ¼-inch Whitworth.

The Vixen dovetail bar is 31 mm in width, but there is another dovetail in widespread use in the US, originally devised for Losmandy mounts, which is 75 mm in width. This system is used by a number of US manufacturers, including Celestron (for their heavy-duty CGE mount), Astro-Physics and Paramount.

Below and on the opposite page is a selection of readily available German equatorial mounts suitable for mounting telescopes of a wide variety of sizes. Motor drives are available extra for the EQ1, EQ2, EQ3-2, EQ5 and Vixen GP, but are built in to the other mounts. You may also have a choice of a tripod or a pedestal to carry the head. Tripods are more convenient for portable instruments, but pedestals have the advantage that they are less likely to foul the telescope as it swings or to get in your way when observing.

Go To systems

The first stage of setting up a Go To telescope is to align it on known objects in the sky, and the way in which this is done varies. The first on the market were Meade, who patented the principle that the scope's "home position" should be level and pointing north, and this remains at the heart of their Autostar system. Other manufacturers therefore had to find other ways of finding the first alignment stars, and you may find that one system suits you better than another.

All systems first require the date, time and location, though with GPS models this is done automatically. For some curious reason many non-GPS models do not even have an inbuilt timer, so you have to reset the time and date whenever you switch on, though they remember their location.

Early versions using the Meade system required the user to first level the scope and then point the tube due north, which is not difficult if you are in

◄ Losmandy mounts, made in the US, are noted for their sturdiness and accuracy. This one, the GM-8, is capable of taking a load of 13.6 kg (30 lb) and includes electronic motor drives with facilities for correcting for backlash and periodic errors in the drive. The axes are stainless steel and it uses needle thrust bearings throughout.

▶ *Four equatorial mounts. L–R: Sky-Watcher EQ2; Celestron CG4; Celestron CG5 (non-Go To, similar to Sky-Watcher EQ5); Sky-Watcher HEQ5.*

the northern hemisphere and can find the Pole Star, but for beginners and from locations where it is hidden this can be difficult. So more recent Meades attempt to do this themselves, using an inbuilt sensor and compass, though with varying degrees of success. The latest version, the ETX LS, even has an inbuilt camera (the LightSwitch) to locate and center the reference stars, promising completely automated alignment.

Celestron's SkyAlign system requires you to choose any three bright stars or planets, which it then compares with the positions of those that should be in the sky at the time. You must drive the telescope to the second and third stars using its motors. They also have alternative and quicker alignment systems which you can use if you do know the names of any of the stars.

Synta's SynScan is used on both equatorial and altazimuth mounts. With the equatorials, you must first set up the mount correctly for your location, in other words with the polar axis pointing north and at the same angle as your latitude. The home position is with the telescope pointing north and uppermost on its mounting. Thereafter, it will find two reference stars of its own choosing for you to center. In the case of the altazimuth version, you need to know the name of one bright star in the sky at the time, which you point the telescope to manually. It will then find its second alignment star for you automatically.

The only systems, therefore, which can be used by a beginner with no knowledge of the sky or the direction of north are the Celestron SkyAlign and the Meade LightSwitch. The Celestron system usually works very well, and is recommended by many users. The LightSwitch system was released while this edition was being prepared and no reviews have appeared at the time of writing. Please visit www.stargazing.org.uk for updates.

— 3 • CHOOSING YOUR TELESCOPE —

There is no ideal telescope for all people. Everyone has their own circumstances that dictate their choice. And many people have more than one instrument to allow for this. So before you plunge in, consider the various factors, including space available, local viewing conditions, your interests in astronomy, and, of course, your budget.

How big?

A telescope can be a big thing. Probably everybody would like the biggest telescope possible, but for most of us this is out of the question. Even if you have the space both indoors and out for a 400 mm (16-inch) telescope, think through the practicalities. An instrument of that size is bulky and heavy, and requires considerable physical effort to set up. Are you prepared to put in that effort every time you want to observe? Anyone with back problems can have difficulty setting up even a basic SCT, and it's a sad fact that many people who retire and buy the telescope of their dreams soon find that they rarely use it because of the sheer discomfort involved. At this point the question of building an observatory arises, which goes beyond the scope of this book.

Little wonder, then, that people often go for smaller telescopes than they would like, simply because of the sheer bulk and space required. Don't underestimate the effort involved in setting up even a common-or-garden 200 mm SCT. These may be regarded as "portable," but in practice "transportable" is more the case. You can't carry the whole thing easily, so you will have to reassemble the instrument every time you want to use it, possibly trampling mud into the carpet as you do so. And by the time you are ready to begin observing, that promising clear evening may have turned cloudy, as UK observers know only too well. Considerations such as these are responsible for the popularity of compact cats that can be up and running within minutes.

But if space is at a premium because you live in a small house or an apartment, do you necessarily have to buy a small telescope? There are penalties to be paid for small apertures – they do considerably restrict what you can observe, and you may feel that the view through a telescope of less than 100 mm aperture is just too dim and disappointing to be acceptable, even if it is quick to set up. Look at the space you have available convenient to your outside door, and think about just how a telescope might fit in. The tube and mounting of a 150 mm reflector can fit neatly and unobtrusively behind an armchair, for instance. And the rewards of using the larger aperture can outweigh the inconvenience of setting it up compared with the smaller instrument. Even a 250 mm Newtonian reflector on equatorial mount need not take up much space, if you have a handy cupboard. The legs come off the pillar, the counterweight slides off the declination axis, which itself unscrews, and the equatorial head can be stored separately from the pillar.

Wherever you store your instrument, make sure it is in a dry and dust-free environment. A zip-up bag for the tube could be a worthwhile investment if it prolongs its life.

Where do you observe?

Linked with the matter of space available is the question of where you will be observing. Apartment-dwellers may only be able to do the minimum of observing from home, so will need to travel to a darker site anyway. Many people do have a garden or backyard, but with observing conditions that are far from ideal, so they, too, will travel to a darker site when they can. Of course, this usually means going by car, though some enterprising types have made their own instruments as large as 450 mm that will fit in a suitcase, so they can travel by air. For most people, however, their telescope will have to fit in a car from time to time, which means restrictions on bulk and weight. If you have no car, you are probably limited to truly portable instruments.

Your interests

Next, consider your interest in astronomy – both the level and the nature. A beginner will probably prefer to get a modest size of instrument, with the hope of upgrading later. This also applies to youngsters – their parents are often keen to nurture their interest, but realize that it might be just a passing phase so they are unwilling to put too much money into the purchase.

Once you are committed, this is still a factor to consider. Someone who simply wants to take the odd look at the planets, or be equipped

► *Lifting an SCT off its tripod is no problem if you are fit, but bear in mind that even this basic Meade LX90 SCT weighs 10 kg (22 lb), while the latest GPS models weigh 19 kg (42 lb).*

◀ Two beginners' telescopes: a 60 mm refractor on equatorial mount at the rear and a Russian TAL-1 110 mm reflector in the foreground. Although the reflector costs considerably more, it gives much more worthwhile views. Its mounting is much heavier and more stable than that of the refractor.

for the occasions when there is something interesting to view, can't justify the same expense or commitment as can a devoted observer.

At the cheaper end of the market, there is a place for the simple 60 mm refractor as a beginner's instrument. It's cheap and will give a quick look at a range of objects. The drawback is that the 60 mm will usually show only a small fraction of the objects in the sky. Unless you have really dark and clear observing conditions, most of the time even after much searching you will simply end up staring at yet another blank bit of sky.

I would recommend even a beginner to get a good 130 mm or larger reflector to start with, if they can afford it – about two or three times the cost of a reasonable 60 mm refractor. This will show a wide range of objects and is still about half the price of a catadioptric of rather smaller aperture.

The nature of your interests may play a very large part in your choice. Someone keen on observing the Moon or planets, or possibly double stars, needs a long-focal-length instrument with maximum contrast of images and a stable mounting. However, extreme accuracy of the motor drive is not too crucial for planetary work as it doesn't matter if the planet drifts a little from the center as long as it remains in the region of best definition of the eyepiece.

Many devoted planetary observers prefer refractors, particularly the apochromatic or fluorite variety, which are expensive compared with long-focus Newtonian reflectors that give virtually as good performance providing they are kept well maintained. Modern apochromatic refractors are far more compact than their traditional counterparts of earlier years, so it is now possible to use a large refractor without requiring a permanent observatory or fixed base. But a large number now use large SCTs, notably the Celestron 9.25 and 14, for planetary imaging using webcams, as described on page 163. The lower contrast of these instruments compared with refractors can be overcome during image processing, and can make good use of even fleeting moments when the seeing is good and the larger aperture reveals more detail.

If I were pressed to choose "the best" telescope – a "telescope for all seasons," as it were – I would probably go for a Newtonian of around 300 to 400 mm aperture, working at about f/8. This is quite a big instrument, and it would have to be mounted on a good, heavy, well-engineered equatorial mount. But we are now talking about an observatory instrument, as this is not the sort of thing that you can easily carry around. The fact remains that a good Newtonian, of reasonably long focal ratio – at least f/6 – takes a lot of beating. It gives excellent images of the planets, and apertures of 200 mm or larger can really get down to some serious deep-sky work. Above all, it is good value for money – you can afford a larger aperture in a Newtonian than in any refractor or SCT.

Its drawbacks are that in the larger sizes it is bulky, and it also requires some care. The reflective coatings on the mirrors collect dust and can tarnish, so they need regular recoating, at some effort and expense.

The basic SCT is adequate for planetary work, as it has a long focal ratio of f/10. Its lower contrast compared with a refractor or a Newtonian reflector may be a drawback, but in its favor there is the possibility of moving it from place to place should your view of the ecliptic, where the planets are to be seen, be obstructed by trees or buildings.

For greater portability you could consider a smaller aperture catadioptric, which some people maintain give better views of the planets than larger apertures – see "Lens vs mirror" on page 41. But always bear in mind that

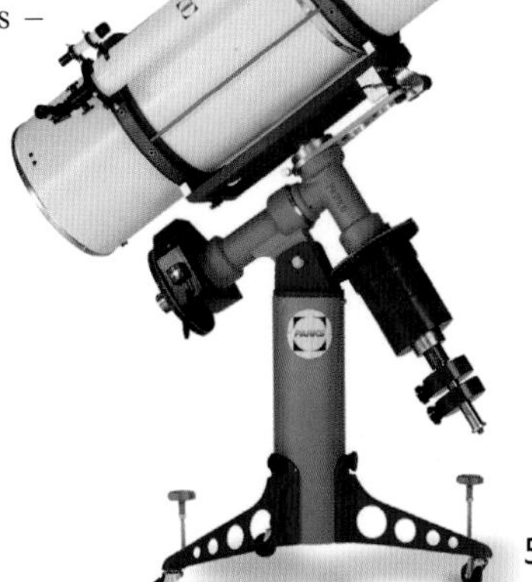

► *Dream telescope? This Parks Optical Observatory 16-inch (400 mm) f/5 Newtonian reflector on a heavyweight equatorial mount is custom-built to order. Such an instrument demands a permanent installation: the counter-weight alone weighs 90 kg (200 lb)!*

with portability and light weight you also usually get lack of stability of the mounting.

A city observer may be forced to observe only the brighter objects, such as the planets. But a country observer, with dark skies, may be more into deep-sky observing, where large aperture is more important than high magnification. A drive may be even less essential, just as it would also be irrelevant to a comet searcher who will need to sweep the skies constantly. For deep-sky work, a large Newtonian or Dobsonian is a good choice.

Anyone keen on deep-sky photography or CCD imaging – the electronic version of photography – would be well advised to place great emphasis on the accuracy of the motor drive, along with the means of constantly monitoring the drive rate. Although many telescopes are now supplied on altazimuth mounts with electronic equatorial tracking (such as the Meade ETX and LX series, and Celestron NexStar), you must have an equatorial wedge for long-exposure photography. For planetary photography, the crucial thing is to have a very stable mounting that will not vibrate if the wind blows.

Lastly, the matter of budget, which for many of us is probably the first thing we think of. But in practice, it probably ranks lower down the scale than you might think. Really, when someone says "I can't afford it," they quite often mean "It's not as high a priority for my money as some other things." For example, many people spend a sizeable fraction of their income on their car. For them, it's a priority expense for ease of getting to work or going shopping. Often they have a larger and more expensive car than they need for these purposes. But if they were truly devoted to observing they could afford a top-of-the-range telescope and would be happy to take the bus to work or run an old jalopy to pay for it.

But given that amateur astronomy is a leisure activity, budget does play a part. There are, however, ways of getting satisfactory telescopes without paying a lot for them. You could, for example, buy a second-hand telescope. Look in the astronomy magazines for these, ask around in astronomy societies or at star parties, and try online websites and forums. You may even be able to find a newsletter devoted exclusively to telescope sales. Some telescope shops also have secondhand telescopes that they have taken in part exchange.

Secondhand telescopes

The matter of buying secondhand telescopes is a tricky one. A telescope that has rarely been used should still perform as well as on the day it was made, as long as it has been stored in dust-free and dry conditions. On the other hand, there are many things that can degrade a telescope's performance, even assuming it was performing well to start with.

Things to watch out for in a secondhand telescope

• Obvious signs of wear and tear on the finish – not necessarily serious.

• Worn or stripped threads on the nuts and bolts holding the mounting together – often replaceable.

• Missing nuts, bolts or washers on such things as the eyepiece holders and the finder assembly. These are virtually impossible to replace with the identical item, so you have to resort to finding something else with the same thread.

• Sloppy gears on the motor drive or focusing mechanism. There may be an adjustment, but if the gear teeth are worn there is little you can do without engineering skills. In the case of an SCT, the work should be done by a specialist.

• Dust on the lens or mirror – can be cleaned (see page 62).

• Tarnished or corroded mirror coatings – will require realuminizing, probably costing a considerable sum.

• In the case of lenses, deterioration of the glass – particularly a problem with the more expensive materials such as fluorite.

• Objective lenses that have been taken to bits in the past and reassembled either wrongly or without the essential spacers – metal shims, usually at a 120-degree spacing around the edge of the objective. But a lack of these spacers does not mean that the lens is wrongly assembled – not all lenses have them.

• In the case of small lenses, usually less than 50 mm aperture, deterioration or mold in the Canada balsam that cements the components of the lens together. They can be cleaned and recemented, but the materials can be hard to obtain and there is no guarantee of success.

• Misaligned optics. Unlikely in the case of a refractor, but very likely in the case of a reflector or an SCT. Most Newtonians are fairly simple to realign (see page 62), but an SCT may require more care.

▶ *The eyepiece end of an SCT has four tiny screws that can easily get lost but are hard to replace. Make sure a used telescope has its full complement.*

All this makes it sound as if a secondhand telescope is not worth buying, but this is far from the truth. If the instrument has been well looked after, or rarely used, there should be few problems. Many secondhand telescopes are underpriced or good value depending on your point of view. An instrument that is in good condition but apart from a few knocks on the finish should give as good views as a new telescope.

Telescope maintenance

Cleaning the optics

The optics are the most crucial part of your telescope. They inevitably collect dust and dirt with use, which degrades performance. But resist the urge to clean every speck – you can do more harm than good by repeated cleaning as there is always the risk of scratching the surface. Some say that you should only clean the optics as a last resort. In all cases, wait until the glass surfaces are completely dry before trying to remove dust.

If you live near the sea, bear in mind that onshore breezes can have a salt content that can be very harmful to the optics and indeed any metal parts of your telescope. Be wary of observing under such circumstances.

Lenses, eyepieces and corrector plates

Use a pressurized air spray or a fine brush, available from camera shops, to remove dust specks. Sometimes the air sprays can themselves leave a residue, particularly if you don't hold the can vertical while spraying. If dust or marks remain and become a major problem, use lens cleaner or a very thin soap solution (just a drop or two in distilled water) in a spray. If absolutely necessary, wipe with household tissues gently from the center to the edge.

Mirrors

Only blow dust away – do not try to clean the mirror on a regular basis. It is better to put up with a bit of dust than risk scratching the reflective coating.

When cleaning becomes inevitable, you will need to remove the optics from their cells. Do not try to disassemble an SCT – this is best left to experts. There is usually no reason, however, why you should not remove the mirrors from a Newtonian. Note how the mirrors are fixed in place and remember that they are always coated on the front surface, unlike household mirrors which are coated on the back so you look through the glass to see the image.

Immerse each mirror in warm, slightly soapy distilled (or previously boiled) water. Photographic wetting agent is better than dishwashing liquid, which may contain other ingredients. Gently sweep the mirror surface with a household tissue under the surface of the water. Rinse it with distilled water with just a drop of wetting agent added (this helps the water to run off and avoids it drying in spots), and allow it to dry in a dust-free area.

When reassembling, take care not to pinch the mirrors in their cells – this can cause strains in the glass, giving rise to poor images. The optics will then need to be recollimated (realigned).

Collimating optics

This is a subject that could run to many pages in order to cover all the eventualities, but only at the risk of getting very boring to most readers. So here is a very potted guide to the subject; a more detailed treatment is available at www.stargazing.org.uk.

Ideally, the alignment should remain fixed, but there is always the risk that a component will move when the telescope is transported, and it is not usually feasible to return the telescope to the manufacturer for recollimation. If there is no handy expert around who has been through it all before, you may have to do the recollimation yourself. Think twice before tackling it, but also bear in mind that poor collimation can completely spoil your view so do not be faint hearted if you are convinced it is necessary.

Every lens or mirror has an optical axis, about which it is symmetrical. The optical axis of each optical component in the telescope should be in line, and making sure this is so is called collimating the optics. In the case of refractors the objective lens should be collimated accurately by its manufacturer, and there is no reason why it should ever go out of line, so this is not a topic that need concern refractor owners. But in the case of reflectors and SCTs there are at least two components, the primary and secondary mirrors and the corrector plate of an SCT, so there is a greater chance of misalignment. In addition, a small shift in a mirror doubles the angle through which light is reflected, making the alignment more critical.

When you look into the telescope through the center of the eyepiece tube, but without an eyepiece present, you should see a symmetrical view, with the main mirror at the outside, the secondary mirror inside that and the reflection of the eyepiece hole dead center within that. Another way of checking that this is so is to defocus a bright star at the center of the field of view. The black hole in the disk that is caused by the outline of the secondary should be dead center.

There are usually adjustments to all the mirror positions, though these may be crude in the case of some Dobsonians. Each mirror generally has three adjusters, at 120° to each other, with maybe a central one for a secondary which you should not undo since it holds it in place. Adjust by trial and error, with only tiny movements at a time as the adjustments are quite sensitive. Do not force the screws – if one is tight, you may need to slacken the other two to allow it to turn.

Start by making a stopper for the eyepiece tube with a hole that restricts your view to the very center of the tube. A plastic 35 mm film canister with a small hole drilled dead center is ideal. In a Newtonian, first adjust the position of the secondary so that it is central in the view and reflects the outline of the main mirror centrally within it. Then adjust the main mirror so that the reflection of the secondary is central within it. Some people put a spot at the center of the main mirror to help with the alignment. This will not affect the optical performance as the center of the mirror is always in the shadow of the secondary.

In the case of an SCT, the only adjustments provided are for the alignment of the secondary mirror, and they are usually hidden behind a plastic plate that prises off or may be held by a screw. The adjustments are very fine and many people believe that they should only be touched by an expert. However, as Celestron's own manual states, "Do not be intimidated to touch up collimation as needed to achieve optimal high-resolution views. It is worth the trouble!"

Many Maksutov telescopes, such as the ETX 90 and 125, cannot easily be collimated by the user.

Recoating the optics

Every so often, the coatings of reflectors will need to be renewed, more often in the case of open-tubed telescopes used regularly near the sea or industrial areas, and rarely for closed-tubed reflectors or SCTs that get only occasional use. In the case of a catadioptric you should return the telescope to your nearest supplier, but with Newtonians and Dobsonians you can remove the optics yourself, pack them securely and send them off for recoating. The normal coating is of aluminum, with an overcoating that prevents it from tarnishing.

Contact telescope suppliers or look in astronomy magazines for details of firms that will do the job for you. Be warned – it is not cheap, so try to look after the optics rather than getting them realuminized.

◀ *The view of a correctly aligned Newtonian, as seen through the open eyepiece tube. Ideally, you should look through a small central hole.*

Although you can check for most of the faults listed by simply looking at the telescope in daylight, the real test is always how it performs on the sky, which can sometimes be hard to arrange. So unless it's so cheap that you can afford to take a chance, try to get that all-important sky test.

You might think that a cheap telescope that underperforms is worth having compared with a smaller but fully functioning telescope costing more. But in my experience a telescope with defects is of very limited use. It has neither convenience of use nor good performance. Quite often, even the low-magnification views are poor and you would see just as much with a pair of binoculars.

Dobsonian telescopes

The great merit of the Dobsonian, and the reason it was designed in the first place, is that it is cheap to make. Invented by San Francisco amateur John Dobson in 1968, the Dobsonian uses simple and readily available materials to keep the optical parts in alignment and provide a simple means of pointing the telescope with just the right amount of friction that it stays where you put it and yet is easy to move.

The classic design has a cardboard tube and pivots that use Teflon

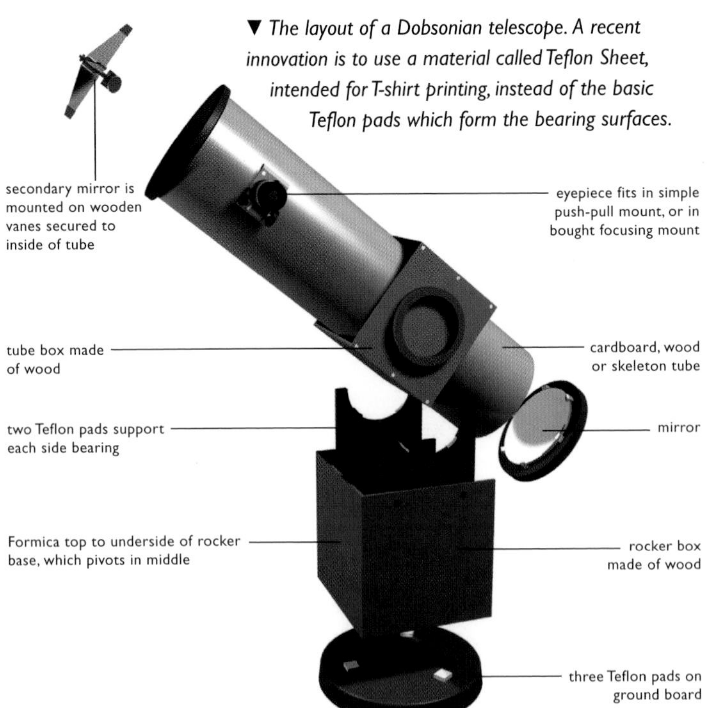

▼ *The layout of a Dobsonian telescope. A recent innovation is to use a material called Teflon Sheet, intended for T-shirt printing, instead of the basic Teflon pads which form the bearing surfaces.*

What if . . .
. . . your telescope is faulty?

Choosing your supplier can be as important as choosing your telescope. Telescopes are very individual items – they can suffer from many more manufacturing variations than, say, a TV set. Some suppliers make their own instruments from scratch, some are retailers of goods supplied by the big companies, while others combine the two.

Most true manufacturers (other than the big companies) will probably not have large stocks of any particular instrument and may well have a considerable backlog of orders, particularly for the larger instruments. So bear this in mind when ordering – indeed, make the delivery time one of your first questions, because there is little point in getting excited over an instrument that could take up to 18 months to materialize. It's best to know that in advance. And mentally double any quoted delivery delay.

If you are buying a mass-produced telescope, you will probably get it from a retailer. The magazines are full of ads from competing suppliers, often at a wide range of prices for the same goods. When choosing, find out whether the firm checks the instruments before sending them out, and also consider what will happen should you discover a fault that requires the telescope to be returned. Don't be swayed by assurances that "We've never had one returned yet." Yours could be the first.

pads bearing against a Formica surface. Many different approaches are possible, but most keep the Teflon pads. Dobson's aim was to provide a design that virtually anybody could make, using only simple hand tools, and it remains the best approach to DIY astronomy. Once you have bought the optics – basically, a main mirror, a flat secondary diagonal mirror and at least one eyepiece – you can make the mounting very cheaply. The most common adaptation is to make

Sky-Watcher Mercury-607

Manufacturer Synta
Type Refractor
Aperture 60 mm
Focal length 700 mm
Focal ratio f/11.66
Mounting Altazimuth on tripod.
Drives etc Manual slow motions only.
Standard accessories 20 mm and 10 mm eyepieces; star diagonal; 5 × 24 finder; 1.5× erecting eyepiece; 2× Barlow.
Weight 4 kg (8.8 lb)
Comments Ideal for occasional use, and as a low-cost starter telescope, but performance and versatility are limited. It is not suited for photography, other than of the Moon.
Alternatives Similar instruments available from most manufacturers.

a square wooden tube rather than use a cardboard tube, particularly when cardboard tubes of the right diameter are hard to come by. Another option is to buy some commercial parts, such as a rack-and-pinion focusing mount for the eyepiece.

Dobsonians come into and go out of favor, and their simplicity can work against them among people who like the high-tech approach of Go To telescopes. But once you have looked at Saturn or a galaxy through a smaller but more costly telescope, then seen the same thing much brighter and clearer through a Dobsonian, you will realize that they knock spots off the competition.

If you go down the Dobsonian route because of cash limitations, you must bear in mind the drawbacks of these telescopes. In the larger sizes particularly they do take up space and are inconvenient to move around. So if you are a cash-strapped student, for example, while the Dob will be cheap to buy, you may find it hard to store in your tiny room and hard to carry to a suitable observing site. Some people actually build two separate mounts, one at home and one at their observing site, and carry the optics from one place to the other.

Case Histories

"We would like to buy a telescope for our 11-year-old who has become interested in astronomy. We think this is an excellent hobby, but we know only too well that it could be a fad. So while we don't mind paying a fair sum for a good instrument, we are cautious about spending too much."

Many parents find themselves in this situation. Few are inclined to buy a top-of-the-range instrument on a "nothing but the best" basis, and

ETX-90

Manufacturer Meade
Type Maksutov-Cassegrain
Aperture 90 mm
Focal length 1250 mm
Focal ratio f/13.8
Mounting Go To fork altazimuth with optional adapter to equatorial; tripod.
Drives etc Push-button slow motions; fast slew.
Standard accessories 26 mm eyepiece; red-dot finder.
Weight 3.5 kg (7.7 lb) plus tripod 5 kg (11 lb).
Comments Lightweight and portable, with good optical quality. Controls (e.g. focus, clamps) fiddly to use. Can be adapted for photography, but small aperture restricts its usefulness.
Alternatives Synta Skymax-90 (UK) or StarMax-90 (Orion US) have same specification but different mounts.

indeed all but the most affluent parent would think twice about spending thousands on a gadget that might only be used a few times.

However, the opposite approach of spending the bare minimum on a cheap department-store refractor would be a big mistake. There's nothing wrong with a basic 60 mm refractor as a beginners' instrument, but there are plenty of bad buys around, even from stores with top names. I have seen rubbish telescopes on sale in some of the top department stores in London, for instance. Any child given one of these would quickly become disappointed by how little they could see.

So buy from a reputable dealer, who knows what they are selling. Having said that, consider buying a telescope one level up from the basic. A good 114 mm reflector has a lot more power than a 60 mm refractor, and is good enough that it will allow youngsters to expand their interest if necessary. At the same time, the resale value should be reasonably good, should you decide to sell off the instrument if your child either loses interest in astronomy, or becomes more interested and you want to get something more powerful.

Bear in mind that in a low-volume market such as that for telescopes, dealer discounts are often high, particularly for the cheaper instruments. In other words, they may put a profit margin of 50 per cent on a cheap telescope, so they are unlikely to want to give you more than half the price back, even if you decide to upgrade just a few weeks later – though this will depend on what you want to upgrade to.

"I have become interested in astronomy late in life and I would love to take it up as a hobby. But ill health means that I don't want to spend long hours

TAL-100RS

Manufacturer Novosibirsk Instrument-Making Plant
Type Achromatic refractor
Aperture 100 mm
Focal length 1000 mm
Focal ratio f/10
Mounting EQ5 German equatorial.
Drives etc None, but can be motorized.
Standard accessories 25 mm and 6.3 mm eyepieces; 6 × 30 finder.
Weight 17 kg (37 lb)

Comments Optically excellent, reportedly with little false color; original TAL-100R is on a basic Russian mount, but now supplied on Synta mounts with Go To option.
Alternatives Celestron, Sky-Watcher and others all have similar instruments.

outside, which is a pity as I live in the country and have room for a good-sized telescope."

There are several possibilities, but if health is a constraint you should think twice about any telescope that requires some effort. For serious hobbying a medium-sized SCT is a good solution, since it is versatile and can be set up fairly quickly. But you should also consider the effort needed to set it up in the first place. The tube assembly is fairly heavy, and many find that they need all their strength to lift it on to its tripod.

One alternative possibility is a smaller computer-controlled catadioptric, such as a Meade ETX or a Celestron NexStar, which most people can carry without difficulty. You can be observing within minutes, and can equally quickly pack up before the cold gets to you. You can observe conveniently seated at a table – also a boon for anyone who is disabled. However, the smaller instruments are rather fiddly to use, so it would be a mistake to aim for too much portability. The single arm of the NexStar 6 makes it comparatively easy to use.

Even a Dobsonian telescope might be suitable. Though they are heavy, they can be put on castors and wheeled out into your observing location, as long as there are no steps to negotiate. You would, however, need to find some way of preventing the base from moving while you observe. As with other Newtonians, the viewing position of a Dobsonian is of variable height, depending on the object you are observing, and the eyepiece is either horizontal or inclined at an angle so that you are looking slightly downward.

Of course, if you can run to a fully fledged observatory you could go for a larger instrument. But you can forget the idea of observing in the

Celestron NexStar 130 SLT

Manufacturer Celestron (Synta)
Type Newtonian reflector
Aperture 130 mm
Focal length 650 mm
Focal ratio f/5
Mounting Single-tine altazimuth Go To with SkyAlign and one- or two-star alignment.
Drives etc Push-button slow motions; variable slew speed.
Standard accessories 25 mm, 9 mm eyepieces; red-dot finder.
Weight 8.16 kg (18 lb)
Comments Well reviewed, with easy-to-operate SkyAlign alignment which makes it ideal for beginners while providing enough aperture for some serious observing of planets and deep-sky objects.
The motor drive allows for webcam photography.
Optically similar to the Explorer 130 PM, but with Go To, at approximately double the cost.
Alternatives Celestron offer refractor versions on the same mount, plus a 114 mm reflector with relay lens in its focuser to shorten the tube length. Sky-Watcher have similar instrument but without SkyAlign.

warm, as the mixing of the warm air with the cold outside air plays havoc with the image quality – what astronomers call the "seeing" (see page 36). The only alternatives are to move to a warm climate, or to use the telescope robotically (remotely).

The latter option is now available even with the smaller computer-controlled instruments, which can be linked to a computer indoors by moderate lengths of cables without needing great expertise. A CCD, digital camera or webcam can also be controlled from the computer, and you can spend most of the time in the warm, though the smaller instruments lack the driving and pointing ability for the best results. Some people even use telescopes set up on a friend's property miles away, far from city lights. But it may require far more involvement than you had envisaged, and it loses the magic of the direct visual view. CCD images may well reveal galaxies and other objects that are far beyond the reach of the eye, but they are just not the same as seeing with your own eye photons that have traveled across a sizeable part of the Universe.

"My husband and I live in a high-rise block with our small child. We are both mad keen on astronomy but are rather strapped for cash and of course we have to drive for an hour or so to get even reasonable skies. Is there any hope for us?"

If you are as keen as you say, you will want a reasonably large telescope. Make a Dobsonian and if necessary keep it in the car! Then the main problem is a domestic one of whether or not your child is happy to accompany you on your observing trips.

If that's a problem, all is not lost for city astronomers. Virtually any telescope, or even binoculars, can be put to good use. You could take

Sky-Watcher Explorer-130

Manufacturer Synta
Type Newtonian reflector
Aperture 130 mm
Focal length 900 mm
Focal ratio f/6.92
Mounting EQ2 Equatorial on tripod.
Drives etc Push-button RA drive; 2× and 8× slew.
Standard accessories 25 mm and 10 mm eyepiece; 2× Barlow; red-dot finder.
Weight 13.75 kg (30 lb)

Comments An excellent performer at a budget price, with 30 per cent more aperture than a 114 mm telescope. The Barlow lens has some false color.
Alternatives Motorized f/5 model also available as 130PM, with parabolized mirror, plus Go To version. Orion US versions known as SpaceProbe have detail differences.

up solar observing, which can be done just as well from the city as from the country and requires only a small telescope, or you could observe variable stars with binoculars. Even the much-derided but cheap 60 mm refractor can be used for variable-star work, though the number of objects you can observe and indeed find is limited.

You may be tempted to buy a small secondhand Go To telescope, but bear in mind that motorized Go To instruments do make quite a noise when slewing across the sky. Your neighbors might find it quite irritating in the small hours if you use the instrument on a balcony close to their windows.

"My career is doing well and I've decided to treat myself to a really good telescope, though I haven't been interested in astronomy for very long. We also have a villa in the sun and it would be nice to be able to take it there as well, which means flying. Should I go for a fluorite refractor or will a computer-controlled SCT be a better bet?"

Many telescope suppliers were surprised by how many people were able to afford the computer-controlled SCTs when they first came on the market. People who have previously never considered themselves to be amateur astronomers snapped them up, impressed by the opportunity of observing without the need for years of apprenticeship learning the sky. Old-timers may sniff, but the extra interest has done the hobby nothing but good.

You probably accept that you are not concerned with dedicated, serious astronomy but are happy just to be able to see things with a good telescope when the fancy takes you. The question is not one of cost,

NexStar 6

Manufacturer Celestron
Type Schmidt-Cassegrain
Aperture 150 mm
Focal length 1500 mm
Focal ratio f/10
Mounting Single-tine fork altazimuth Go To on tripod; long-exposure photography possible using optional equatorial wedge.
Drives etc Push-button slow motions; nine speeds. SkyAlign plus four other alignment options.
Standard accessories 25 mm eyepiece; red-dot pointer.
Weight 13.6 kg (30 lb)
Comments Single arm gives easier access to focuser than conventional fork mount. Cannot be used without electric drives. NexStar 4 and 5 have built-in equatorial wedges for astrophotography.
Alternatives Meade ETX LS – claims completely automated alignment procedure.

but of convenience and indeed fun. The most expensive telescope isn't necessarily the best for everyone. They generally bring added complexity, which could make you reluctant to bother with observing. The golden rule is that the best telescope is the one you use the most. So consider two telescopes: maybe a fluorite refractor at home on Go To mount for taking a look at the Moon and planets from your light-polluted suburb, and a Dobsonian in your villa for those glorious dark nights when you can just let your view wander through the Milky Way at random. However, if you want to go down the SCT route, and are prepared to ship the instrument from place to place, you could certainly have hours of fun putting it through its paces.

Many people take their SCTs with them on board aircraft. You always run the risk of the baggage-handling system sending it to Bangkok (unless that is your destination), but the alignment of the optics should withstand all but the most extreme handling.

Refractors are regarded as being more sturdy than any SCT or reflector, but in practice there should be little to choose between them.

"My husband is just coming up to a special birthday and I want to get him a telescope as he is a great fan of Star Trek. But I don't know the first thing about telescopes. What should I do?"

It might be worth discussing your situation with a specialist dealer who will take into account the various other factors such as cost, portability and so on. This may be a good case for getting a 70 mm refractor, possibly with computer control, as a low-cost fun telescope. If you want to spend more, you might choose between a larger

Meade LXD75 SN8

Manufacturer Meade
Type Schmidt-Newtonian catadioptric
Aperture 203 mm
Focal length 812 mm
Focal ratio f/4
Mounting Meade LXD75 equatorial on tripod.
Drives etc Meade Autostar Go To.
Standard accessories 26 mm eyepiece; 8 × 50 finder.
Weight 42 kg (92 lb)

Comments Short-focus reflector on equatorial mount, more suited to astrophotography and deep-sky than planetary observing. Rather lightweight tripod.
Alternatives No other Schmidt-Newtonians available, but conventional Newtonian from Sky-Watcher. Also available in 152 mm and 254 mm aperture.

Newtonian reflector or one of the neat little Maksutovs, both of which cost about the same. You may be faced with a choice of either a Meade ETX-90 with Autostar computer control, or a conventional 200 mm (8-inch) Newtonian on equatorial mount, both of which are comparable in price. The ETX is fun, automatic and portable; the Newtonian will show much more, but is cumbersome and won't find objects for you. It's a fair guess that the fun scope will be better for someone who is not already a committed amateur astronomer, but be aware of the choice.

"I already own a small telescope but find that it doesn't really show me much on the planets and most deep-sky objects are scarcely visible. I'd like a larger telescope but I wonder if it's worth it – will I just see more faint fuzzy objects and a bit more detail on the planets? I have the room and the money for a large telescope, but although I live in a country area there's still an increasing amount of light pollution, and I fear that it would be a waste of money getting a telescope these days. So should I take the plunge, or what?"

There are several issues here: what will a large telescope show; is there any point, given increased light pollution; and which telescope is the best for you?

What is called "aperture fever" struck amateur astronomy in the 1980s and 1990s, with a trend toward sizes that just a few years ago would have definitely been in the "small professional" category. The rise of the Dobsonian has made it possible to get telescopes around half-a-meter (say 20 inches) in aperture into the family car, though these are usually purpose-built. The major manufacturers offer telescopes up to 400 mm (16 inches) aperture in either Dobsonian, Newtonian or

Tele Vue-102

Manufacturer Tele Vue
Type Apochromatic refractor
Aperture 102 mm
Focal length 880 mm
Focal ratio f/8.6
Mounting OTA only; shown here on Gibraltar altazimuth mount.
Drives etc n/a
Standard accessories n/a
Weight 4.1 kg (9 lb); tripod 7.7 kg (17 lb).

Comments A classic apochromatic refractor, sold as tube only but with all accessories available including tripod.
Alternatives Other manufacturers offering apochromats of this size include Stellarvue, Vixen and Takahashi.

Schmidt-Cassegrain formats. So while a 300 mm (12-inch) telescope was once regarded as large for amateurs, these days many star parties feature instruments considerably larger.

Such large apertures are wasted on planets, at least for most of the time. The major limitation on seeing planetary detail is our own atmosphere, whose turbulence gives rise to blurred, unsteady images – what astronomers call bad seeing (see page 36 for more details). In fact, you can often see more detail on the planets with a smaller, rather than a larger aperture. There is an optimum size of around 200–300 mm, for general-purpose instruments at least, above which there is little gain when observing the planets.

But the main question is "How much can you see anyway?" The sad fact is that no telescope will show the dramatic and beautiful colors in nebulae, or the scintillating spiral arms of galaxies, that show up in photographs. There is only the vaguest hint of color in the Orion Nebula even with large amateur telescopes, and the spiral arms of even the brightest galaxies are only ever faintly visible. The reason is that a photograph or CCD image builds up light over many minutes, and can record color in objects that are too faint for the eye to see. There is a basic problem of optics that makes it impossible to squeeze enough light into the eye to bring out the true colors of objects that are not points of light. The drawing on page 9 gives a good idea of what you will actually see, given good conditions, though even this had to be drawn more boldly than it appears in reality, in order to show up at all.

As for light pollution, you can judge for yourself. If the Milky Way is clearly visible at times of year when it is high in the sky, your site can't

LX90

Manufacturer Meade
Type ACF Schmidt-Cassegrain catadioptric
Aperture 203 mm
Focal length 2000 mm
Focal ratio f/10
Mounting Go To fork altazimuth with optional adapter to equatorial; tripod.
Drives etc Full Go To capability with GPS, periodic error control.
Standard accessories 26 mm eyepiece; 8 × 50 finder; red-dot pointer.
Weight 23 kg (51 lb)
Comments Claimed to be the best-selling serious telescope in the world, this is a good-value package with a high degree of functionality. Suitable for all forms of imaging with appropriate autoguiding CCD or guidescope.
Alternatives Celestron CPC 800 GPS; more advanced SCTs from Meade and Celestron.

be too bad. If it is only visible on rare occasions, however, a large telescope may prove to be wasted.

Assuming, however, that your skies are good then there is no reason why you should not enjoy the use of a large telescope. Get the largest you can afford, taking into account your secondary interests. If you want to do photography or CCD work, then pay great attention to the quality of the mounting. If, however, visual observing is all you want to do, a Dobsonian may be all you need. It's even possible to equip a Dobsonian with a device that will help you find objects in the sky. Although it will not drive the telescope around the sky to find the object, it will indicate on a handset which way to move the telescope to find the object you have chosen from a computer database.

But if the idea of viewing faint fuzzy objects does not appeal, forget about a telescope and log on to the Internet instead. The Hubble Space Telescope site has plenty of pretty pictures.

"I have been an avid amateur photographer all my life and now have gone over to digital. I see amazing photos in the astronomy magazines, taken with digital cameras and quite ordinary telescopes, but is it as easy as all that? My experiments years ago with photographing through my 76 mm refractor were a big disappointment, but it looks as though with today's gear I should be able to do much better. What's the secret?"

The ads for many telescopes make it sound as if astrophotography these days, whether with film or digital devices, is a simple matter. But while the apparatus is a great deal better, from the telescope quality to the mounts and drives, it will never be as simple as taking a snap during

SkyQuest™ XT12 IntelliScope®

Manufacturer Orion/Synta
Type Dobsonian
Aperture 305 mm
Focal length 1500 mm
Focal ratio f/4.9
Mounting Dobsonian with digital readouts (push-to).
Drives etc None
Standard accessories 25 mm and 10 mm eyepieces; 9 × 50 finder.
Weight 37 kg (83 lb)
Comments Enough aperture for most purposes, and with electronic finding system for about the same cost as a 125 mm Go To Mak yet lighter than a 250 mm SCT. Unsuitable for general photography.
Alternatives Portable truss version (XX12) also available from Orion as well as smaller models without digital readouts; similar Dobs from several major manufacturers.

the day. You need to pay close attention to every aspect of the process to get the best results – the optics and their collimation, the focusing, guiding during a time exposure, and subsequent processing. Although digital cameras and CCDs are more sensitive than film during long exposures, the time taken for each successful result can be considerable. There are often hours of work at the computer after you have taken the shots. So don't assume that the electronics does the job for you.

Undoubtedly the area of astrophotography where it's now possible to get good results without too much outlay is webcam photography of objects in the Solar System. Webcams are cheap, they can be focused fairly quickly because they give a video output in real time, and they can give dramatic results quite speedily. Even a good-quality driven telescope as small as 80 mm will give webcam results which rival the best that could be obtained by amateurs using even large telescopes and film a generation ago. So hone your skills with this cost-effective setup, for which virtually any driven telescope is adequate.

For long-exposure work the size of telescope is not too critical, though for small and faint objects the larger the better, but the ability to guide it accurately is essential. An off-the-shelf SCT on Go To mounting is adequate, but for good results you will need a secondary guide telescope on the same mounting, with a CCD autoguider (see page 173). The mechanics inside the fork mountings of some mass-produced SCTs are notoriously unreliable, however, and a German mount is generally reckoned to be superior for astrophotography. As for cameras, digital cameras are adequate if you have good skies, but for best results cooled CCD cameras are needed (see page 164).

SPX250f6.3

Manufacturer Orion Optics
Type Newtonian reflector
Aperture 250 mm
Focal length 1600 mm
Focal ratio f/6.3
Mounting Go To German equatorial on pillar.
Drives etc Starmap display on handset provides sophisticated pointing system with variable speed slewing.
Standard accessories 25 mm and 10 mm eyepieces; 50 mm finder; Crayford focuser.
Weight 29 kg (64 lb)
Comments A very versatile and durable system, capable of top-quality results for all observing purposes, and for imaging with the appropriate off-axis guider or guidescope at lower cost than an SCT of the same aperture. Not readily portable.
Alternatives Entry-level reflectors of same aperture from Orion Optics, Synta and other sources are considerably cheaper but less solid and without Go To.

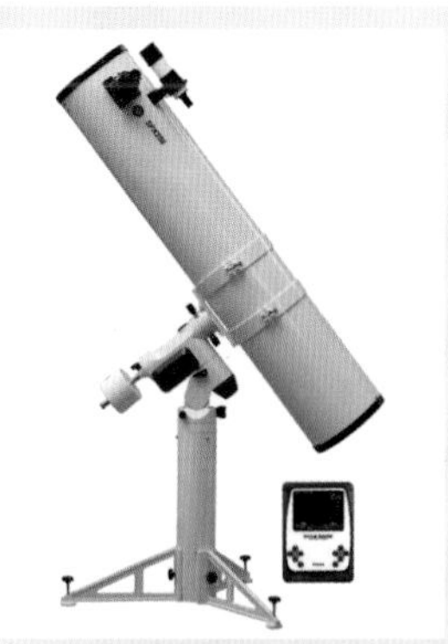

"Because I live in a badly light-polluted suburb I have hesitated to get any telescope at all, let alone a decent one. What can I observe from such a location? I have heard that there are light-pollution filters. Will they help?"

The Sun, Moon and planets, double stars, variable stars . . . virtually anything except deep-sky objects is well within your reach. Don't let the sky brightness put you off buying a good telescope. You may be restricted to a smaller range of objects, but many of them actually change from time to time, whereas a galaxy will look the same in 10 years' time as it does now.

For solar observing you don't even need a particularly large telescope. If you get bitten by the solar bug, you might consider a special solar telescope that will show you prominences and amazing detail on the Sun's surface that changes daily (see page 154). For the rest of the Solar System, a 150 mm telescope is excellent for observing most of the planets, though a larger one will give brighter views. In the presence of light pollution, a long-focus instrument is an advantage.

If you are really badly light-polluted, filters won't make a lot of difference. They are not capable of cutting out the white light that is making an increasing contribution to the suburban sky, so don't expect too much from them whatever your telescope.

In practice, even through light pollution a larger telescope is capable of showing deep-sky objects that a smaller telescope would fail to show. If you are keen on CCD work, you can photograph quite faint deep-sky objects that could not have been seen visually from your location for many decades because of the sky brightness. Image processing can work wonders that even the best LPR filters cannot hope to achieve.

CGEM-1100

Manufacturer Celestron
Type Schmidt-Cassegrain
Aperture 279 mm
Focal length 2800 mm
Focal ratio f/10
Mounting Go To German equatorial with GPS on tripod.
Drives etc Full Go To operation with large database.
Standard accessories 40 mm eyepiece; 9 × 50 finder; star diagonal.
Weight 54.4 kg (120 lb)
Comments At the limit of transportability. Suitable for a wide range of observing and imaging with the appropriate guiding systems.
Alternatives Meade 250 mm or 300 mm LX200. Larger SCTs also available.

4 • STEPS TOWARD FIRST LIGHT

When you buy something new, you look forward to using it with great anticipation. The first drive in your new car; the first CD played on your new hi-fi; that new computer. And the first clear night after you have bought your new telescope should be full of delight. However, there is a steep learning curve associated with using a telescope, and unless you know what you are doing, the first night's viewing can be disappointing.

Simply looking through a telescope is a skill that needs to be practised. Then there is the matter of actually finding objects to look at. It is one thing to have lists of interesting objects in the sky, such as those on pages 182–7, but quite another to find them. This is something of a chicken-and-egg situation: you can't find objects until you have some practice using the telescope, but you can't get that practice until you can find the objects. It really does help to get experience with the telescope in daylight, by looking at whatever objects you please – trees, houses and so on.

Even if you have an all-singing, all-dancing computerized telescope that finds the objects for you, unless you know how to observe you may well decide that your telescope is not working properly. By taking that short-cut across the learning curve, you might be missing out on valuable experience.

First, familiarize yourself with the telescope itself. Telescopes may vary widely in appearance, but most have the same basic features, as shown in the box "What's what" on page 82. Some telescope handbooks are not as helpful as they might be, so I have tried to cover all the options.

Let's assume that you have a basic non-motorized, non-computerized telescope. For the time being it can be absolutely any telescope, of any aperture, on any kind of mounting, as long as it has that essential requirement – a finder telescope.

Many people would say that a telescope is only as good as its finder. The job of a finder is to give a low-magnification, wide-field view of the same part of the sky as the main telescope, with a crosswire in the field of view, to help you aim the main telescope at your chosen object. Finders are specified in the same way as binoculars – that is, by magnification and aperture in millimeters. Many finders supplied with telescopes are basic 6 × 30 types, which are adequate in their own way but do not have a particularly wide field of view.

A few more expensive telescopes have 8 × 50 finders or similar, with a field of view similar to binoculars. It is not uncommon to see telescopes equipped with home-brewed finders made from the optics from old binoculars. Making and inserting a crosswire can be tricky, so bear this in mind if you are tempted to try it. For more on finders, see Chapter 7.

At the other end of the scale, the very cheapest telescopes have tiny finders which are supposedly 5 × 24. Some of these are not even true 24 mm telescopes – there is a stop behind the lens which reduces the effective aperture in order to sharpen up the image a bit. Their image is so dim that they show little more than the brighter stars, which makes the job of finding faint objects particularly difficult.

Many telescopes these days are supplied with a red-dot finder instead of an optical (telescope-type) finder (see also page 157). These shine a red dot on to a clear plastic screen, the idea being that you peer through this from a short distance behind it and view the

▼ *Many supposedly 24 mm finders are stopped down to 10 mm. The stop is visible through the objective (top). Sometimes the stop is easily removed (bottom) but this makes the performance of the finder even worse.*

Testing your telescope

At some stage, most people will want to be sure that their telescope is working as it should. There are two reasons for this: you may want to check the quality of manufacture, or you may suspect that something has changed and want to find out what is going wrong. Either way you should carry out the same tests, which fortunately can be done without dismantling anything.

Before testing you must be sure that the telescope has been outside long enough to reach the night-time temperature but with no dew on the optics. Never expect to get good results pointing it through an open window, or even through a closed one. You will also need a night when the atmosphere is steady, which may mean waiting for the right occasion. The first test is designed to check this point.

Begin by locating a brightish star of about second magnitude using an eyepiece of moderately high power – say about 150. If your telescope has no drive, use Polaris (in the northern hemisphere) which will stay in the field of view for a long time. Defocus the star slightly, moving from one side of focus to the other. It will expand to a disk of light. Any movement within this disk when you are outside the focus point suggests that either the atmosphere is turbulent or there are air currents within the telescope, in which case you will have to try again later or on another occasion.

The defocused disk of light should appear virtually the same on either side of the focus point – circular and even. If there are spikes or corners on the disk, something is wrong. These can be caused by a temperature imbalance, but it is more likely that there is a fault in the optics. The most common is that the main mirror is being pinched within its cell, particularly if there are three deformities to the disk at 120 degrees to one another.

Check that the eyepiece is not to blame by rotating it in its mount. If the error rotates with the eyepiece, the fault lies there and not with the rest of the telescope.

Next, check the evenness of the disk. There should be rings within it, but you should not find that the outside edge is bright on one side of focus and faint on the other side. This also indicates a fault with the optics. Any general unevenness could be caused by an overall poor finish.

If the disk shrinks to a short line or oval on one side of focus and another oval or line at right angles on the other side of focus, there is a fault called astigmatism somewhere in the system.

In the case of a reflector or SCT, enlarge the disk by defocusing until the shadow of the secondary mirror appears within it. If the test star is in the middle of the field of view, this shadow should be in the center of the disk. If it isn't, the optics are misaligned. Correct the misalignment if you can by following the instructions on page 62 before worrying any further.

This brief guide can only tell you how to check if something seems to be wrong, and not how to identify the source or what to do about it. But if there seems to be a problem, try to get a second opinion from someone more experienced before complaining to the manufacturer or returning your telescope. Double check on another occasion, preferably with another instrument of similar size available.

red dot against the sky – an illuminated peepsight with no actual magnification.

Red-dot finders are now cheaper to make than optical finders and some people prefer them, but they have several drawbacks. It is often hard to see the stars behind the red dot, and there is a knack to positioning your eye in the right place in the dark. In light-polluted skies they are of much less use than an optical finder that shows stars fainter than you can see with the naked eye. And, worst of all, if you forget to switch them off, the tiny battery runs down and they are useless.

Whatever type or size of finder you have, its job is to help you aim the main instrument, so it follows that the two must be aligned perfectly with each other. This is a job that is best done by day, as ground-based objects are easier to distinguish and don't have a habit of moving through the sky. But you must choose a test object as far away as possible. If you want to know how far away, consider the separation between your finder and the centerline of the telescope. If this is, say, 20 cm then the finder will be looking 20 cm to one side of the main telescope. Astronomical objects are at infinity, so this doesn't matter. But imagine that you are looking at a chimney pot a few hundred meters away. Through the main telescope the whole thing is quite large in the field of view, and 20 cm is quite a large proportion of the image. You could allow for the difference by guesswork, but it is better to choose an object so distant that 20 cm is negligible. But do not, under any circumstances, aim at the Sun or any ground-based object near the Sun.

If you are aligning the telescope at night, it is better to choose an object near the celestial pole (see pages 101 and 102 for how to find this), as this will not move very quickly though the sky. But if you are unfamiliar with using the telescope, trying to find a star without the aid of the finder in the first place is tricky.

▼ If you are aligning on a nearby object (left), allow for the separation between your finder and the tube – but remember that the view is inverted.

When adjusting the finder (right), gently turn two opposing screws simultaneously in opposite directions so as to keep the tube trapped between them.

To align the finder, first find an object in the main telescope using the lowest magnification you have – use the eyepiece with the longest focal length, if you have a choice. Remember that objects will appear upside down in the field of view unless you are using a star diagonal. Squint along the top of the tube to see roughly where the telescope is aimed. Once you have found an easily recognized object in the telescope, clamp it and use the slow motion controls, if you can, to make sure that the chosen object is dead central in the field of view.

▲ *If the finder is loose within the outer ring of its mount (top) wrap tape around it until it is a tight fit (bottom).*

Now go to the finder and loosen the little screws that hold it in place. Look through, but don't be tempted to move the main telescope at all. Instead, adjust the finder's position until the same object is in the field of view. Now comes the fiddly task of tightening up the screws so that the chosen object is dead center on the crosswires. The trick is to slacken one and tighten another simultaneously, so that the telescope is held between the two as it moves.

Some finders on budget telescopes have only one set of adjustment screws. In this case, the finder should be a tight fit in the short tube that holds it, so that there is something to adjust against. But the finders often do not even touch the tube so there is nothing to adjust against. In this case, wrap self-adhesive tape around the tube until it is a proper fit and you will find that adjusting the finder suddenly becomes a lot easier.

Check from time to time that your exertions in moving the finder haven't dislodged the object from the center of the main eyepiece. Once the finder is aligned precisely on the same thing, tighten it firmly and lock it using the locknuts, if any. Double check by moving the telescope to an object in a completely different direction, to make sure that nothing shifts as the telescope swings around the sky. Time spent on this will be well repaid at night when it comes to finding objects in the dark.

First light at last

Eventually comes the big moment – first light! Daytime objects don't really count. Only light that has crossed space is deemed worthy of entering an astronomical telescope.

What's what

Telescopes are really quite simple beasts, and most have basic components in common. You should be able to identify the following:

• **Focusing mount.** This takes the eyepieces and in most telescopes except Schmidt-Cassegrains provides the focusing. There is usually a small screw to hold the eyepieces in place. You focus by means of the knurled knobs on either side, or sometimes by twisting the whole eyepiece assembly. Adjusting the eyepiece farther from the lens or mirror, toward you, focuses on nearer objects.

In SCTs, the focusing knob is on the back plate of the telescope. The eyepiece mount remains stationary while you focus, as the knob moves the mirror itself, inside the tube.

• **Finder telescope.** This is a small low-power telescope which is pretty well essential for locating objects as it has a wide field of view compared with the main instrument.

• **Finder adjustment screws.** You use these to make sure that the finder is properly aligned with the view as seen through the main telescope.

• **Axes.** There are two axes for moving the telescope. In an altazimuth these are called the azimuth (sideways) and altitude (up and down) axes, while on an equatorial mount they are the polar (or RA) axis and the declination axis. The polar axis should be set up pointing north in the northern hemisphere, and south in the southern hemisphere, at the same angle as your latitude.

• **Axis clamps.** Each axis has a clamp to prevent the telescope from moving off the object you are viewing. On motorized telescopes the polar axis clamp locks the telescope on to the drive, so it will then follow objects through the sky if properly set up. The declination axis may also have a drive. It is important to make sure that both axis clamps are free before trying to direct the telescope at an object, otherwise you risk stripping the drive gears. At the same time, when clamping the telescope, do not use more force than necessary to avoid damage to the clamp surfaces.

• **Slow motions.** These are manual geared drives, one on each axis, that allow you to move the telescope slowly to center on or follow an object. They only work when the telescope is clamped on the appropriate axis.

• **Cradles.** These hold the telescope tube to the mount in reflectors and larger refractors. Quite often you can rotate the telescope in its cradles so as to bring the eyepiece to a convenient observing position, and often you can slide the telescope up and down within the cradles to balance it.

• **Counterweight.** Only needed on German-type equatorial mounts. Make sure it is correctly positioned on its shaft to balance the telescope before use.

• **Motor drive controls.** There may be a separate on-off switch on the base, then the polar and declination drives (if fitted) are controlled using the buttons on the handset. There may be buttons for slewing the telescope across the sky at a faster rate, and for switching from northern to southern hemisphere directions.

• **Wedge.** Only found on fork mounts, for converting them from altazimuth to equatorial. There is a scale of degrees for setting them to the correct angle.

• **Axis adjusters.** These allow you to adjust the mounting when setting it up accurately as an equatorial.

• **Polar telescope.** Aligned with the polar axis, for quick alignment on the celestial pole. There may be two scales used when setting this up – one for a longitude adjustment to suit your location within your time zone, and one for the time and date. More details on setting up are given on page 93.

• **Setting circles.** Graduated scales of right ascension and declination – sky coordinates – to help you find objects by their position only. These only work when the telescope is aligned equatorially.

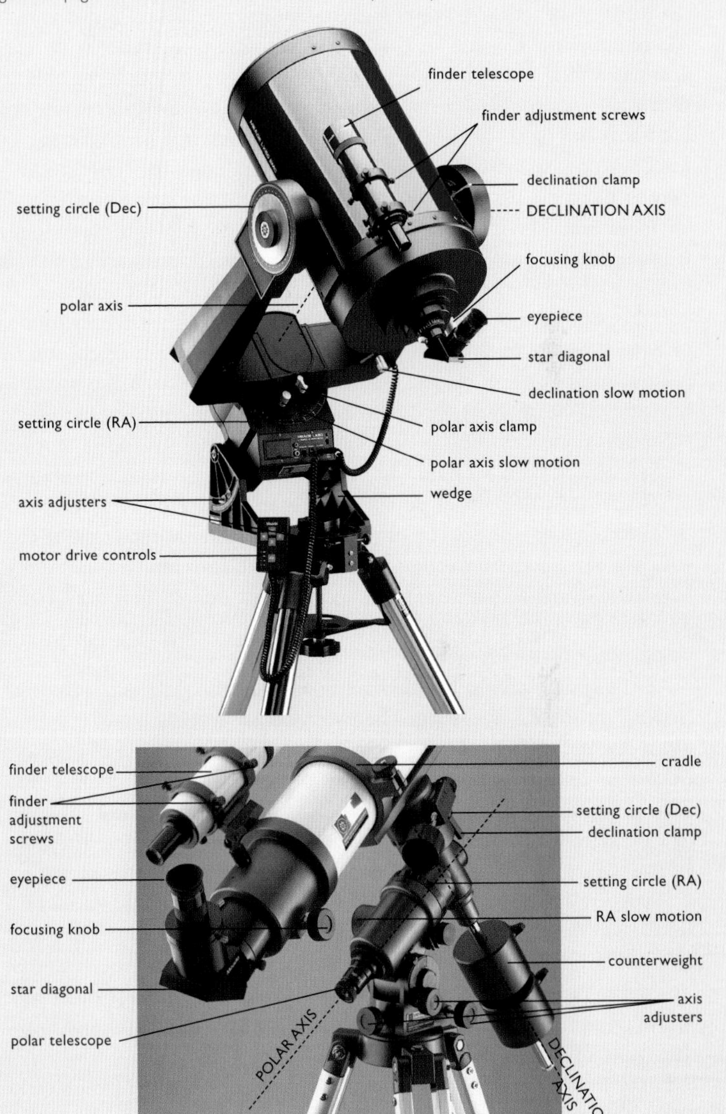

Choose an eyepiece that gives a low magnification (often called low power) to begin with. This generally means using the longest focal length – typically 25 mm or 26 mm. Slip the eyepiece into the fitting, tighten up the locking screws and remove the cover, if any, from the telescope's mirror or lens (no marks for forgetting this step!).

Probably the best object for your first look is the Moon, if it's around. There are many reasons for this. It's hard to miss, so even if your telescope is a little out of alignment with the finder you should be able to find it with no trouble – just aim the telescope in the direction where the light gets brightest. Even if the drive of the telescope isn't working or the telescope isn't driven, it stays in view for a long time. The Moon is easy to focus on, which is very helpful if you are not sure just what you are looking at. There's no doubt when you have the Moon in sharp focus – a fantastic landscape of craters and mountains springs into view. And finally, it's bright enough to penetrate any amount of light pollution.

If the Moon is not around, choose a planet instead, preferably Jupiter or Saturn which are large and unmistakable. Or try for a bright star or cluster, such as the Pleiades, but whatever you choose, make it something that is clearly brighter than the other objects in that part of the sky.

Squint along the tube to get the telescope roughly aimed at your chosen object, then look through the finder with the telescope clamps off. With luck, the object you have chosen should be right there in the field of view. It should be obvious whether you have the right thing, which is why you have to choose something which is by itself in the sky. Move the whole telescope slightly and see how the object moves. Once the object is centered nicely on the crosswires, look through the main eyepiece for your first view.

At this point you may discover that there is a certain knack to looking through a telescope. The eyepiece has a fairly small eye lens, just a few millimeters across, and you have to get the pupil of your eye exactly in line with it. Remove your glasses if you wear them – you should be able to focus the telescope to allow for most cases of short or long sight. You should only need to use glasses if you wear them for an eye defect such as astigmatism rather than for long or short sight.

If you are not observing the Moon, there is something to be said for starting to observe when there is still some light in the sky. That way, you will see a disk of light through the eyepiece, and will be sure that you are in line with it. If I seem to be laboring this point, it is because I have given many people their first view through a telescope, and know that quite often they have difficulty.

Focusing is usually easy enough – turn the focusing knob slowly. Probably the telescope will be out of focus when you first look, though it would be a good idea to leave the focus position where it was during daylight. The fewer things you have to do at night, the better. This is particularly true in the case of a catadioptric telescope, which has a wide focusing range with no indication as to the actual focus setting, unlike a refractor or reflector with a rack-and-pinion focuser in which the focus setting is more visible.

Except in extreme cases of defocusing, at low power there is usually something visible. An out-of-focus star will appear as a featureless disk, which in the case of a reflector or an SCT may have a central black hole caused by the central obstruction. As you approach the focus point this disk will become smaller until at focus it is a point of light. Then as you turn the focusing knob further, it starts to expand again.

Having found your chosen object at lower power, you may want to increase the magnification, which you do by replacing the eyepiece by one with a shorter focal length. Before you do this, make sure that the object is exactly at the center of the field of view, and that the new eyepiece is ready to use. Swap them over and the object should still be close to the center of the field of view, but it will probably be defocused, in which case you will have to refocus.

This is where your troubles can begin. If you have an undriven telescope, you will find that when you increase the power, the object moves through the field of view much more quickly, so you have to move the telescope much more frequently to follow it. And with any telescope, not only is the field of view magnified, but also some objects are dimmer. If you are lucky, you will see a blur which you can easily refocus. But often you see precisely nothing, which means that either the object you were looking at is no longer central, or that it is so defocused that you can't even see it.

If a quick turn of the focuser does not help, you may be tempted to move the telescope a little to try to refind it, but usually it is better to put the lower power back in and recenter the object, or in the case of an undriven telescope allow for its drift to bring it back dead center. If the same thing happens again, the new eyepiece is probably considerably out of focus. In this case, try aiming at a really bright star and swapping the eyepieces so that you stand the best chance of seeing the defocused image. Then learn how much you have to turn the focuser, and in which direction, to correct the focus between the two eyepieces.

This is particularly important in the case of small catadioptric telescopes, which can require a considerable amount of turning of the focuser. A small elastic band may work wonders here. By day, when you can see what you are doing and the object doesn't move, experiment by moving each eyepiece in and out of the focusing tube to focus rather than using the focuser. Probably only a few millimeters of movement is enough. Slip an elastic band around the barrel of each at the point where it is in focus. It will now be a simple matter to swap eyepieces without refocusing. Such eyepieces are termed parfocal. Metal parfocalizing rings are available if you want to replace the rubber bands with something more permanent.

If the atmospheric conditions are not very steady – and they rarely are – at very high powers the object itself may appear to be in motion. Under really bad conditions (called poor seeing) a planet will be rippling and jumping, and may even appear as several overlapping images. A star looks like a writhing spider, with a blurred central body that has numerous fleeting extensions.

If you are horrified by the prospect that all your wonderful new telescope can show is a view that looks as if it is seen from underwater, don't panic. There are often solutions for bad seeing, which can have three causes – general atmospheric conditions, the local microclimate at your observing site, and thermal instability within the telescope. Simply waiting for the telescope to reach night temperature can do the trick – though cover the optics, to prevent the formation of dew. As the night wears on, local sources of heat may cool down, including such things as foliage and paved areas. Even moving the telescope around the observing site might help, if your line of sight to an object crosses some local problem area such as a chimney, rooftop or boiler vent. But many nights, often the clearest and most sparkly ones, have unconquerable bad seeing. All you can do on these occasions is look at objects that don't need high magnifications, such as large nebulae, open star clusters and galaxies.

If you have an equatorial motor-driven but non-Go To telescope, you will need to get into a little routine to find and view an object. First, with the RA drive and declination axis unclamped, locate the object in the finder. Next, with a low-power eyepiece locate it in the main telescope. If you have lined up the finder properly, it should be dead center in the field of view. Once you have found it, tighten the declination clamp straight away. If the object drifts out of the field of view, you should now be able to bring it back in by moving only the RA axis. Even if the telescope is only approximately aligned correctly, this should work. Having confirmed that the object is still in the field of view, operate the drive clutch so that the motor now drives the instrument. In the case of telescopes with manual slow motions, the same procedure applies except that you will need to turn the RA slow-motion cable to keep the object in the field of view rather than relying on a motor.

Don't forget to unclamp both axes before you move to another object or you will damage the axes over a period of time. It is also bad practice to compensate for any imbalance, maybe caused by adding a heavy eyepiece or a camera, by overtightening the clamps.

Advanced observing tips

The actual business of observing needs practice, but here are some pointers which could improve your viewing skills from the start.

Use both eyes

Rather than close their non-viewing eye, experienced observers generally keep both eyes open when observing. It is easy enough to concentrate only on the telescopic view as long as the surroundings are dark. But while this rule for viewing comfort was fine in the 19th century when the surroundings were dark, many people find that their non-viewing eye is now staring at a scene which is much brighter than the object under study. So if you are searching for faint objects, consider using an eye patch.

Get dark adapted first

The eye adapts to dark conditions in two ways. First, the pupil expands to its maximum size within a second or so, letting in more light. The maximum aperture of your pupils varies with age. When you are young it can be as large as 9 or 10 mm, but by your 50s it is probably down to about 5 or 6 mm.

▼ *The* SkyMap *computer program, like others intended for use at the telescope, has a "night" mode that turns the display red to avoid ruining your dark adaptation.*

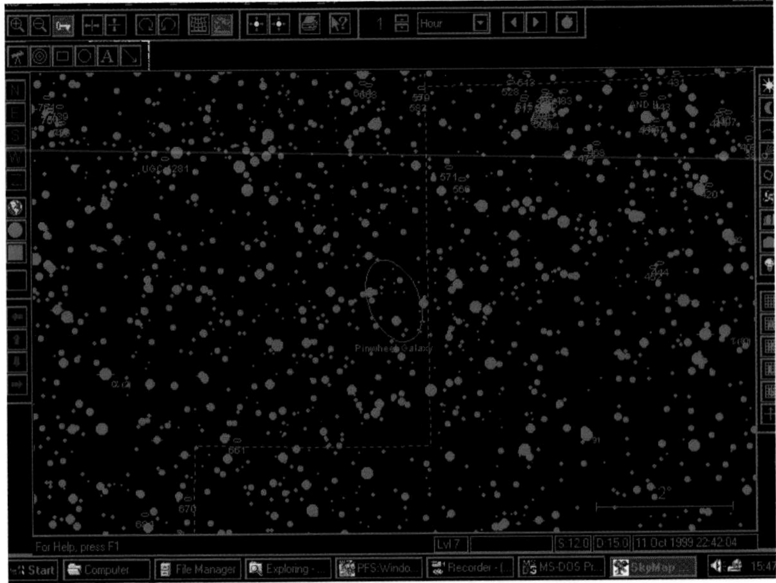

But a more important change takes place more slowly – the production within the eye of a chemical called rhodopsin, or visual purple. This greatly improves your night vision, but it can take up to half an hour for the full effects to be felt. Visual purple is destroyed by bright lights, particularly those with a strong blue or ultraviolet content. This category includes fluorescent lights, computer monitors and TV screens. So avoid these in particular before observing. The bane of all observers are those camping lights with a battery-powered fluorescent tube. Non-astronomers visiting you for a night's observing often bring these along, and helpfully switch them on when you are searching for an eyepiece, which then brings the session to a halt while you all try to recover your night vision.

Tungsten lighting is less troublesome, and red light has the least effect. Astronomers always use red-covered torches or flashlights, and the red LED observing light is a popular accessory.

If you use a computer program to plan your night's observing either reduce the screen brightness before you go out or, if it is a purpose-designed astronomy program, look for a "night" setting, which displays everything in red. You can also customize the Windows colors to shades of red and yellow. The result may not be aesthetic but it can make it much easier to observe afterward.

In many houses the back door leads from the kitchen, which often has fluorescent lighting. Keep this light off when observing, and provide an alternative, such as a table lamp, for other members of the household who may need to use it.

Avoid outside lights

The astronomer these days is plagued by outside lighting. Even if your observing site is not directly illuminated by a streetlight, there are still the house lights of other houses and, worse still, the dreaded "security lights," ruefully dubbed "insecurity lights" by many amateur astronomers. For a trifling sum, householders can buy a kilowatt lamp which will be visible for miles, which they use to illuminate their garden and the surrounding neighborhood all night. These and other lights in your vision can ruin your observing, so try to screen them from your sight, maybe using a temporary method such as a blanket thrown over a stepladder which you can position to block off a troublesome light.

Keep optics clean

Astronomical objects are often dim and low-contrast by their very nature, so give them the best chance of being observed by keeping the optics of your telescope clean. Cover mirrors and lenses with a dust-tight cover when they are not in use, but if they become dewed up, let the dew evaporate before putting it on. This may mean first using a makeshift loose-fitting cover such as a circle of cardboard, to allow air to circulate, before putting on the proper one. A dew-covered mirror or lens is at risk from any airborne dust, which can then become stuck to its surface as the dew dries. The business of cleaning and recoating optics is covered on page 62.

Eyepieces, too, should be kept clean and not carried around in a jingling heap in your pocket.

Use averted vision

When trying to see a faint object, don't look directly at it but look to one side of where you believe it to be – the technique of averted vision. It is actually easier to spot faint objects when they are not at your center of vision. Your eye has two types of light-detecting sensors, called rods and cones. The cones are responsible for color vision, and they are closely packed at the center of vision, while the rods, which are more sparsely scattered over the rest of the field of view, give monochromatic vision but are more sensitive to light. So using the rods improves your chances of seeing an object, although you can see less detail.

While on the subject of the eye, bear in mind that you have a blind spot some distance from your center of vision, away from the nose in each eye. An object at this point may appear to wink on and off as the small "saccadic" eye movements, which take place all the time when we look at a view, bring it on and off the blind spot.

When you are straining to see an object at the limit of vision all sorts of odd effects occur. The view might appear somehow grainy, for example, even though you know this is an illusion. Some people say that the eye can store up light to a small extent, like photographic film, so it can also help to stare fixedly at one spot to reduce the saccadic movements and therefore keep the faint object on the same rods in your eye so the light builds up. All I can say is, try it and see if it works for you.

Tap the telescope

If you are looking for a faint object, it can help to tap the telescope slightly. It's often easier to see a slightly jiggling object than a steady one, particularly by averted vision. Our peripheral field of vision is very sensitive to small movements, probably because this had good survival value in the days when our ancestors were trying to avoid being the hunted rather than the hunter. This suggestion is of course at odds with the previous one.

Odd objects

When you are looking at a bright object, such as the Moon, you may see objects moving across its surface. Occasionally these are high-altitude objects such as migrating birds, insects or weather balloons, but they may also be "floaters" in your eye. These are normal and are an inconvenience, but if eye defects are troublesome you may be getting advance warning of something more serious, so consult your doctor.

Don't overmagnify

Beginners sometimes want to increase the magnification excessively in order to see more detail. But there is a limit to the amount of detail any telescope will show, set by the size of the lens or mirror. So once you have made this detail clearly visible to the eye there is no point in increasing the magnification further, in theory at any rate.

As an example, take a 100 mm (4-inch) telescope. The finest detail this can show, according to both theory and practical tests, is just over 1 arc second. The ideal human eye can see detail as fine as 1 arc minute, so in theory we should only have to magnify the image 60 times to be able to see all the detail this telescope can show. In practice, not everyone can see such fine detail, and it would be worth allowing a factor of three or four for ease of observing. So a magnification of 180 to 240 should show all the detail that can be seen with the instrument, even under ideal viewing conditions.

Turning this into a general rule gives a high-power limit of twice the aperture in millimeters (or 50 per inch). And in the case of larger telescopes, the limit is often set by the seeing, although there are usually odd moments of steady air which allow you to see much more detail than usual.

▼ The Moon as seen in the northern hemisphere (a) with the naked eye or binoculars; (b) directly through a reflector or refractor; (c) using a star diagonal on the telescope. In each case, north and east (as seen in the sky) are marked. The rotation of the sky makes all objects move from east to west, as shown by the arrow.

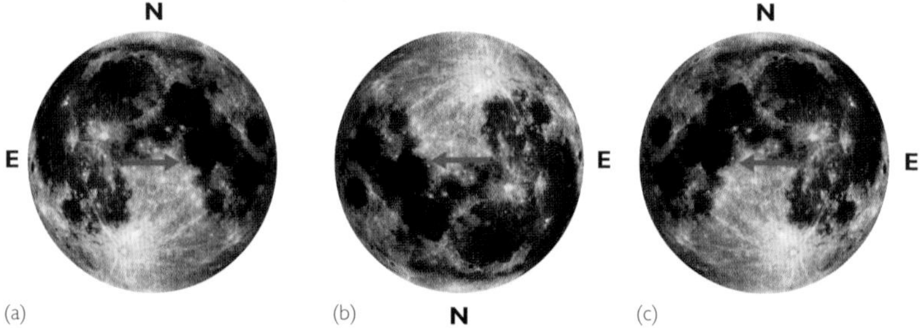

What if...

...you don't see anything when you look through?

The view appears totally black.

• Have you removed the lens or mirror cover?

• Check that you are using the lowest-power eyepiece – that is, with the longest focal length.

• Remove the eyepiece and look into the instrument while shining a light into the main aperture. It is possible that a component has become misaligned, in which case it is best to leave things alone until you get the telescope into the light.

There is some light visible.

• Try focusing. Even bright objects can disappear if they are wildly out of focus.

• Has the telescope been knocked, so that the finder is no longer aligned? Try slowly sweeping from side to side and up and down to find the object.

• Has one of the optical surfaces become dewed up? If so, don't wipe it. Use a hair dryer if possible, or simply cover up the optics loosely and leave them for a while. It makes sense to keep the telescope covers in a warm spot while you are observing, which will help the dew to evaporate quickly. The worst thing to do is to leave them face up so they also collect dew. Chapter 7 describes accessories for avoiding dew.

Get your bearings

Virtually all astronomical telescopes give an upside down image. This is because they tend to use the simplest optical system, which inverts the image. Light is precious to astronomers, the argument runs, so there is no point in adding extra optical components to bring the image the right way up, called an erect image. Actually, in this day and age of high-quality multicoated eyepieces this is no longer very important, but the attitude seems to be that real astronomers don't need erect images.

The inverted image can lead to problems, however. It matters little when simply viewing, since it is not usually necessary to relate what you see to the naked-eye view. But it is most obvious when viewing the Moon, since you will often be comparing what you see with a lunar map. It also becomes evident when you move an undriven telescope to follow an object through the sky: it moves in the opposite direction to what you would expect, though this becomes second nature very quickly. Most finders also give an inverted image, so there should be no discrepancy between the image in the finder and that in the main instrument.

Things become more complicated, however, when you add a "star diagonal" to a refractor or catadioptric. These are invariably included with the telescope to avoid the need to crouch down on the ground when observing, and in the case of short-focus, fork-mounted instruments such as SCTs are essential for many viewing angles. Star diagonals laterally reverse the image, so the Moon, for example, is back to front, though it is now the same way up as in the sky.

To get your bearings in such cases – to find out which is east and west, north and south – begin by allowing an object to drift through the field of view, switching off the drive if necessary. Objects always move from east to west, which in an inverted view is from right to left in the northern hemisphere. Astronomers sometimes refer to this movement by using the terms "preceding" and "following." A star "preceding" a brighter one will be leading it through the sky, so will be to its west.

The positions of north and south depend on whether you are using a star diagonal. The simplest way to be sure is to move the telescope slightly toward the pole, looking to see which way the image shifts. Without a star diagonal, the lower part of the field of view should be nearer the pole (north or south, depending on your hemisphere).

Probably the most difficult inversion to cope with is that caused by the finder. Most astronomical finders give an inverted image, like the main field of view. You just have to get used to the fact that the starfield you are looking at through the finder is inverted compared with the star map and the sky.

This mental inversion, however, should be quite unnecessary. It is of little benefit to see the finder image inverted in the same way as the main telescope view. Many observers agree, yet manufacturers still make most finders of the inverting type. If you have trouble using your finder, it may be worth trying to find a non-inverting finder (see page 157).

The final thing to check before observing is that if your telescope is on an equatorial mount it is properly balanced. With both axes unclamped it should move freely, but it should stay put when you stop pushing it. If not, make sure that the tube is balanced by moving it up and down within its cradles, and that the counterweight is balancing the tube. You should also roughly align the mount north–south and at the correct angle (see page 93), an operation that should take you only a few seconds once you know what you are doing.

Beyond first light

Of all the things that worry newcomers to observing, setting up an equatorial mount and using the setting circles come top. Many equatorial mounts these days have polar alignment telescopes that are meant to simplify the process, but for many people these just make matters worse.

Setting up an equatorial mount

The fact is that for most general observing you can get away with very approximate alignment of the equatorial mount. There is no doubt that it makes observing easier, though if you have a fork-mounted telescope that can just as easily be used as an altazimuth, you can observe without bothering with the alignment at all. But you will find that you have to keep shifting the telescope, whereas once it is set up even roughly

Problems with Go To telescopes

A Go To telescope should make life easier, but quite often your first sessions are frustrating because the instruction manuals miss things out (and who reads manuals anyway?). So here are some tips for getting your Go To going.

• Don't worry too much about the precision of the leveling or the initial information you give the telescope. It needs this only to find the first two reference stars, and to decide which objects are above your horizon at the time of observation. Once you have aligned on the first two stars, the database should know how to find all other objects. However, if your Go To telescope has no finder, as with the ETX-80, locating the reference stars in the main instrument can be very time-consuming, so in this case precision can be important.

• When using the Celestron SkyAlign system, avoid choosing bright stars that are in a straight line, and aim for widely separated stars.

• If the ETX handbox buttons do not appear to be working, it may be because the speed setting is too low. Press 6 or 7 on the number pad to increase the slew speed.

• "Train" or calibrate the telescope motors, as described in the manual, as soon as you can. This allows them to take into account the characteristics of the motors. Training will need to be repeated if you use a different ETX handbox.

• Telescopes list all the objects in the database, whether or not they are visible from your location at the time of observation. So decide in advance what you want to observe rather than picking at random from the instrument's database.

• Synchronize the telescope to the database regularly if you can. On an ETX, use the handbox to find an object, press "Enter" for two seconds, center the object precisely in the main instrument and then press "Enter" again. This updates the computer's alignment.

• When finding faint or small objects, first synchronize or realign the instrument on a known and easily found object, such as a named star, as close as possible to the faint object. This will improve the pointing accuracy.

• Do not rely on the initial setting of the altitude/declination scale of an ETX. It is easily dislodged.

• On an ETX, take care not to overtighten the altitude/declination clamp, which requires only firm hand-tightening. The azimuth/RA clamp, however, should be tightened as much as possible to avoid slippage.

you can carry on viewing without touching the telescope at all. This is a great benefit, because every time you touch a small telescope it usually shudders a bit, which wastes observing time.

The basic alignment of an equatorial mount is that the polar axis should point north (or, in the southern hemisphere, south) and at the same angle to the horizon as your latitude. There is often a scale on the side of the mount to help you get this angle right. And as long as the mount points within five degrees or so of the right direction, objects will stay put in the field of view for minutes at a time when using a driven mounting, which makes observing much easier.

But in order to use the setting circles, or to take long-exposure photographs, you need more precise alignment. To use the circles you should be as accurate on the pole as a fraction of the field of view of your main eyepiece. Say your eyepiece has a field of view

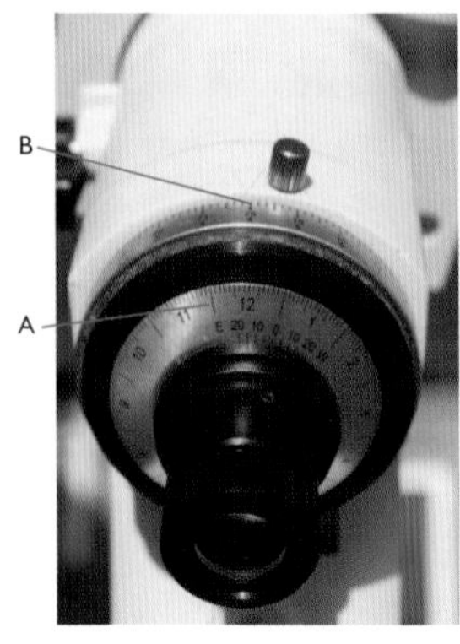

▲ *The scales on a Sky-Watcher HEQ5 polar alignment telescope allow you to correct for your location within your time zone (A) and for the date and time (B).*

of 40 arc minutes (that is, two-thirds of a degree). If you are within a third of a degree of the pole, and your setting circles are accurate, objects should be just within the field of view, though they may be right on the edge. Long-exposure photography is more critical. If you have a telescope of 2000 mm focal length and you want to take 10-minute exposures, you should really get the axis of the telescope within 1 arc minute of the pole.

Fortunately, this isn't too hard to achieve with modern mountings that have polar alignment scopes. These look directly through the hollow polar axis and have graduated markings in the field of view, though older instruments simply had a "bore scope" which looked along the axis, with no markings. You had to estimate where the pole should be. Others show a graduated scale, as shown below, with no further settings.

To use a modern polar alignment scope, you must first set a scale that corrects for your location within your time zone. This is not as straightforward as it might be, because there are numerous variations in time zones to allow for local geography. But in general, the center of your time zone will be a multiple of 15 degrees longitude east or west of Greenwich. So if you are at longitude 88 degrees, for example, near Chicago, you are just two degrees east of the center of your time zone, at 90 degrees west, which is six hours later

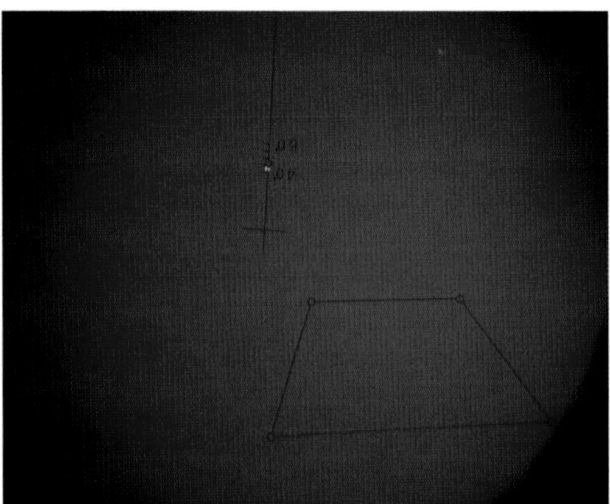

◀ *The view through a polar alignment telescope. The figures 40′ and 60′ refer to the distance of Polaris in arc minutes from the true pole, marked by the cross. In 2005 Polaris was 42.5 arc minutes away; it is 40 arc minutes away in 2017. The other marks show the positions of stars near the south celestial pole, the nearest to the pole being Sigma Octantis, at magnitude 5.5.*

than Greenwich time. This time difference does not bother you, but the main thing is that you should set the scale to 2 degrees east.

Next, you set the date and time on the adjacent scale. The months may be numbered 1 to 12 and the hours 18 to 6. If you are setting up in winter before 6 pm you just have to guess where the mark is! Now you are ready to squint through the alignment scope. A little red light is provided to show the graticules (markings) inside the scope. There are two sets of markings: one shows the location of Polaris by a circle, for use in the northern hemisphere, and the other shows fainter stars near the south celestial pole (SCP). Finding Polaris is no real problem: it is indicated by the Pointers (see the diagram on page 100) and you know when you have it because there is another, fainter star a short distance away. But the stars round the SCP are more elusive, particularly in a light-polluted area. Once you have found them, never let them go – that is, try to make sure you can position the mount in the same place again, at the same angle. Incidentally, it is not essential that mounts are leveled before you begin the polar alignment, though little bubble levels are often incorporated in the mount. As long as the polar axis is parallel with the Earth's, you can have the mount on sloping ground, if necessary.

Some polar alignment scopes have no scales for getting the alignment precisely correct. They simply have rotatable markings as shown in the diagram (A) below. Notice that there are three short arcs that show the offset of Polaris from the true pole in different years, here 1990, 2000 and 2010, though the correction is slight and your telescope may have just a single marking here.

Begin by centering Polaris, then look in the sky for the rest of Ursa Minor, notably Kochab (shown in red in the diagram). Turn the alignment scope so the offset Polaris markings are at the same angle as

▼ *The three steps to polar alignment using a polefinder scope without time scales. The method is explained in the text above.*

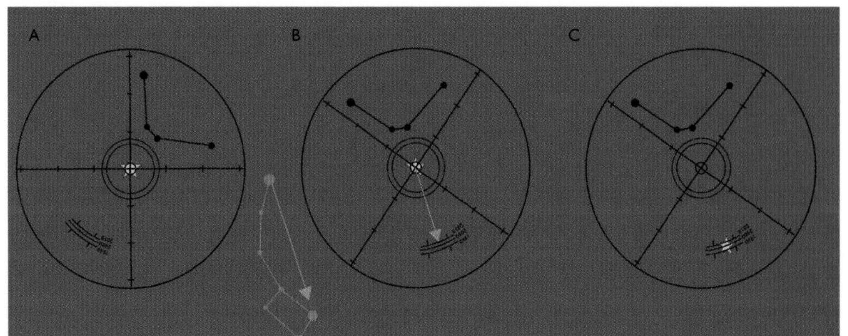

Kochab in the sky, with Polaris still in the center (B). Finally, adjust the whole mount to move Polaris to the center of the correct arc for the year (C). In the southern hemisphere, you will have to locate the other pattern of stars linked by lines and line them up precisely – a tricky task.

These instructions are fine for mountings with polar alignment scopes, but what about the many mounts that have no alignment system at all? Begin by getting as close to the correct alignment north–south as you can using simple methods, such as using a map, a compass or the position of Polaris. Use a protractor or the scale provided to set the polar axis to the same angle to the horizon as your latitude. This should be good enough for general observing, but for greater accuracy, such as for photography, you need to make the observations detailed below using a high-power eyepiece with crosshairs. Anyone who will be doing long-exposure photography will need one of these anyway, for guiding the exposure (see page 167).

Point the telescope at a star near the meridian and near the celestial equator and follow it for a period of time – up to as long as the sort of exposure times you expect to be carrying out. If the star drifts poleward (in either hemisphere), the mount is pointing too far to the west of north, and if it drifts away from the pole it is pointing too far east. So shift the mount a little bit and try again until the star remains central as you follow it.

Now find another star, also near the equator, but near the western horizon rather than on the meridian. This time, a poleward drift means that the polar axis is too low and vice versa. Adjust

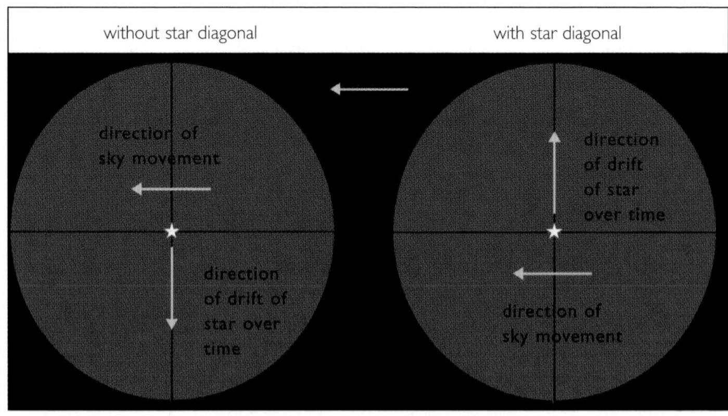

▲ *The direction of drift of a star as seen through a telescope with a normal inverted image without and with a star diagonal. In each case, a poleward drift over a period of time is shown.*

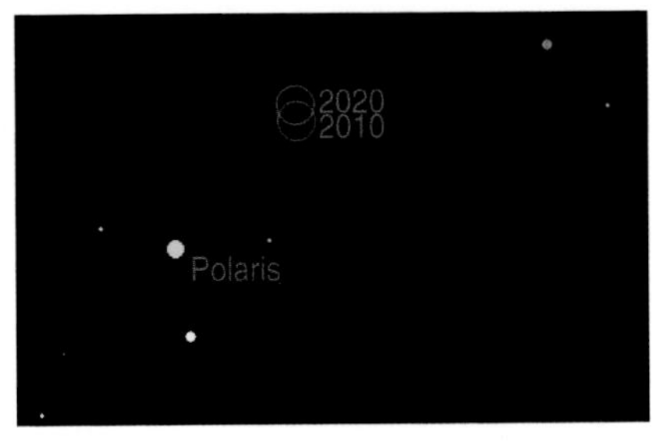

▶ *The region near Polaris, as seen in a finder, with the location of the true pole in 2010 and 2020 shown by circles of 8 arc minutes diameter. Use the nearby sixth-magnitude stars to estimate its position.*

and try again. Should your western horizon be blocked, you can choose a star to the east but now a poleward drift means that the polar axis is pointing too high.

It can take an hour or two of drifting, checking and resetting before everything is just right, so it is not a task you want to perform every time you observe. There are other ways of checking the polar alignment, but this one is generally reckoned to be the simplest.

There are a few proprietary pole-finding attachments for SCT fork mounts, but one alternative is to set the telescope's declination to 90° and use the telescope's own finder to locate the true pole, using the map shown above.

Using the setting circles

This is probably the thing that confuses beginners more than any other. Now here's a secret. It confuses plenty of advanced users as well. They would prefer to find objects by star hopping, or even simply casting around in the hope of finding what they are looking for, than by using the setting circles. You would imagine that circles should make life easier – just dial up the quoted position of a star, and there it is. In practice, there's more to it, and it doesn't help that there are several ways of arranging setting circles.

The declination circle is the easiest to get to grips with, and is usually straightforward. As long as the telescope is even approximately aligned with the pole, you can check the working of yours fairly easily. It is graduated in degrees from +90 degrees down to 0 degrees and then back to −90 degrees. Point the telescope at the pole, and it should read 90 degrees. Point it at a star on the celestial equator, and it should read 0 degrees. In fact, point it at any star and it should read the declination of that star because declinations in the sky are linked to their distance from the pole. Be sure to use reasonably accurate declinations, that is, given for an epoch close to that of your observation and not 1950 (see page 108 to find out about sky positions and why the date makes a difference).

If the circle is slightly in error, then either your mount is not accurately aligned on the pole, the telescope is not accurately placed on the mount (in other words, the optical axis or centerline of the optics is misaligned with the declination axis) or the circles are badly applied. There is not usually any way of adjusting the circles, except on ETX telescopes whose declination scale is all too easily moved, so it may be simplest to remember that in all cases your circles read, say, a degree and a quarter high. Most setting circles are not particularly finely divided: a half or quarter degree is about the closest you can get (see "Reading setting circles" on the opposite page).

Just using the declination circle, you can now find objects with any known declination. So next you need to look at the RA scale, which is graduated in 24 hours with subdivisions. However, unlike the declination scale, which is fixed in place, the RA scale can usually be rotated. This is because the position of a particular right ascension is continually on the move as the Earth turns. Ideally, the scale should therefore be driven at the same rate as the polar axis, if there is a drive fitted, so that once you have set it by choosing a star of known RA, it will then stay set for as long as the drive is kept running. Unfortunately, this is not always the case and some RA circles must be reset to the position of a star of known RA before each use.

Some circles, such as those fitted to some Russian telescopes, actually run the wrong way. I have never understood why this should be so, as it seems most perverse, but to use them you have first to find a star of known RA, calculate the difference between that and the RA of the object you want to find, and dial up the difference. No wonder many people give up on setting circles altogether!

▲ *The declination circle of a Celestron SCT. The telescope is currently pointed at declination 47 degrees. This scale can be read to an accuracy no greater than one degree.*

If you are observing in the southern hemisphere, however, the RA circle runs the opposite way from in the northern, so the Russian scopes are OK there. Some instruments have two sets of numbers on their RA scales to allow for this, so you must choose the correct scale for your hemisphere.

Circles can be very useful when trying to locate faint objects, particularly in today's light-polluted skies. Just trying to find a suitable star in the area of the Virgo Cluster as a starting point for star-hopping can be a

Reading setting circles

Setting circles are usually marked only every degree or two degrees of declination, and every ten minutes of RA. It is not too difficult to judge finer settings, and there may be a vernier scale to help. This is used in place of a single marker point, and consists of a line of markings labeled 0 to 10 or similar. You will notice that these markings do not exactly coincide with those on the main scale. Typically, there are ten markings on the vernier scale for every nine on the main scale.

You can, if you wish, use the 0 marking as the marker point. But to get finer precision, see which line on the vernier lines up exactly with a line on the main scale. This indicates the subdivision you want, read off the vernier scale rather than the main one. Here are some examples using the RA and Dec scales on a Vixen mount.

RA scale

(A) This reads exactly 7h 30m, as shown by the zero marking on the vernier scale.

(B) This is somewhere around 7h 35m – but is it 35, 36 or 37 minutes? The markings for 6m line up precisely, so it is actually 7h 36m.

Dec scale

(C) A declination of 44 degrees exactly, read from the zero line on the vernier.

(D) Now the reading is between 44 and 46 degrees. The 1½ marking lines up with one on the main scale, so you add 1½ degrees to the 44, giving 45½ degrees. You read off the value on the vernier and add it to that shown by the zero line of the vernier – don't pay any attention to the actual marking on the main scale (in this case 54 degrees).

A

B

C

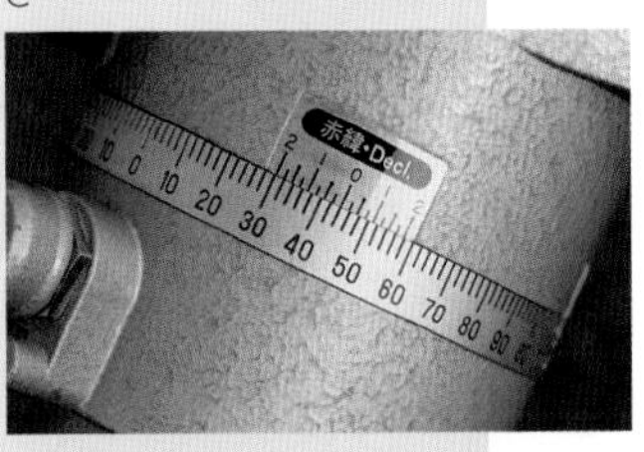

D

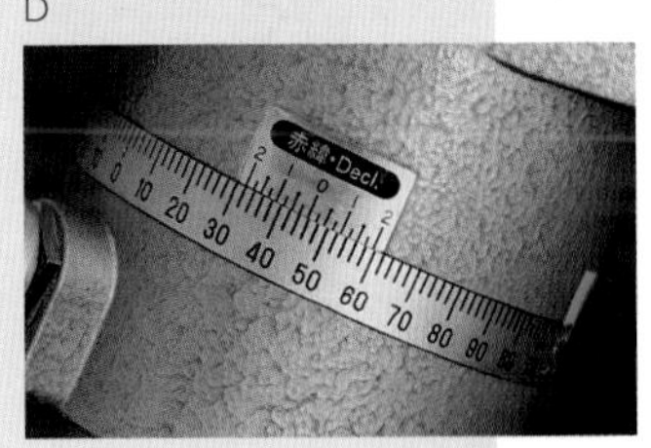

challenge. It's not surprising that more and more people are giving up the battle altogether and are selling their souls to computer-controlled telescopes that make life so much easier. But there is something to be said for using the traditional methods of star-hopping and setting circles: you do get to know the sky. There is real satisfaction in finding a deep-sky object simply by grabbing hold of a telescope and pointing it in the right direction. And you don't need any batteries.

5 • FINDING YOUR WAY

One of the biggest hurdles that the beginner faces is finding all those tantalizing objects that are referred to in books.

Starting on page 182 there is a list of objects that you can look for with various sizes of telescope, but to the beginner the locations given are perplexing. How can you tell which are visible tonight, and how do you actually find them?

This is where owners of the latest computer-controlled telescopes may feel smug. In theory, you just set up the telescope on two bright stars, tap in the name or number of the object you want to observe, and off goes the telescope to find it – though quite possibly it will either

▼ A view of the whole sky from mid-northern latitudes. This is the view you see looking directly upward, so east and west are reversed compared with a map of the ground. The sky rotates around the north celestial pole, which is indicated roughly by following the two Pointer stars of Ursa Major. The constellation of Leo rises in the east, reaches its highest point due south, then sets in the west.

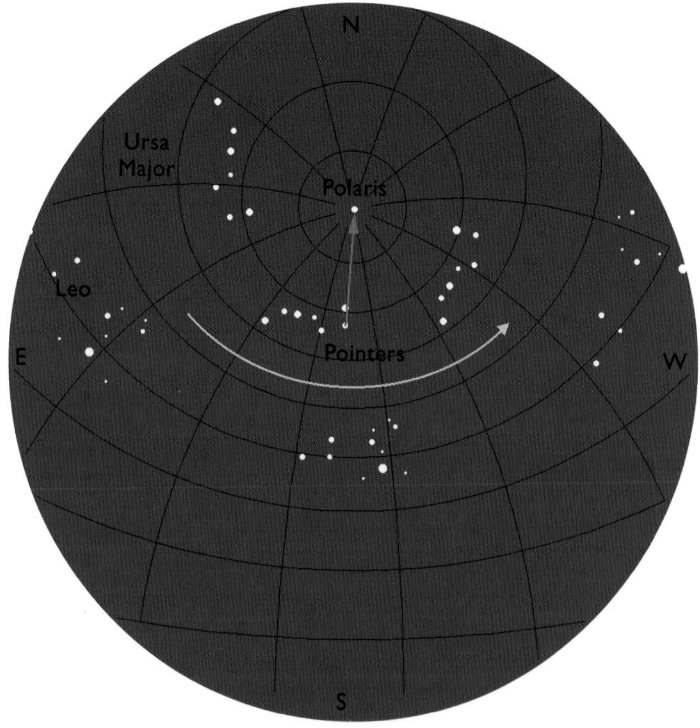

warn you that the object you want is below the horizon or if it does find it, all you see is a blank bit of sky, or maybe a tree that's in the way. And you still need a knowledge of your whereabouts to set up the telescope to find anything at all, as well as some idea of what you want to look for.

There's no substitute for gaining even a basic knowledge of the sky. Only then will you know what's up and what's not, and whether it's worth looking in the first place. You will be able to tell at a glance if a particular object is visible, behind a house or simply too faint to be visible in the prevailing conditions.

Learning the sky isn't something that can be put across easily in a book. Like any other practical skill, there's no substitute for going out and doing it for yourself. But there are tips and tricks to finding your way that will turn all those stars into familiar friends. Once you know where you are, you can be observing within minutes if you go about it in the right way.

To start with, get your bearings – that is, know where north or south are from your observing location. If there's any doubt, remember that the Sun is more or less on your north–south line at midday. This line is called your meridian, which is a useful term as it applies wherever you are on Earth. In the northern hemisphere the Sun is more or less due south at midday, while in the southern hemisphere it is due north, but in both cases it is on your meridian (but see the box "What's the time?" on page 114 for more details of this).

Now you can get to know the way the sky moves. Of course, it is the Earth turning from west to east that makes the Sun and other objects in the sky appear to move from east to west, but from the observer's point of view it's the sky that moves. At night, after a period of observing, you can see the overall movement of the star patterns. In fact, the whole sky appears to pivot around the celestial pole, which is directly above the Earth's pole. This is an essential point to know when getting your bearings. It is located due north (or south in the southern hemisphere) at the same angle above the horizon as your latitude. Northern-hemisphere observers are fortunate in that by chance there is a bright star, Polaris, very

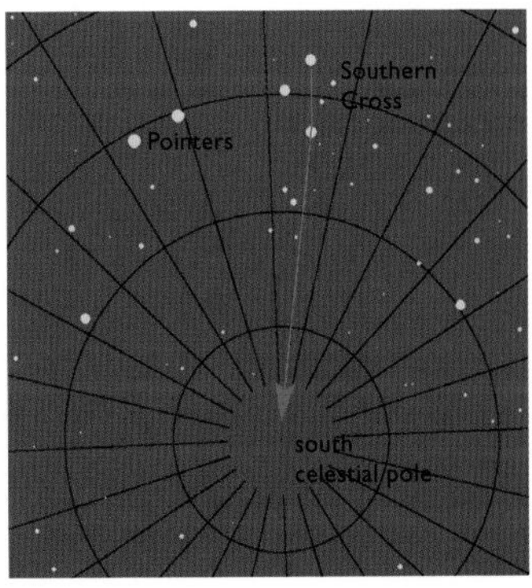

▶ *The southern hemisphere's Pointers indicate the Southern Cross, which in turn approximately points to the south celestial pole. Project the line $4\frac{1}{2}$ cross-lengths southward.*

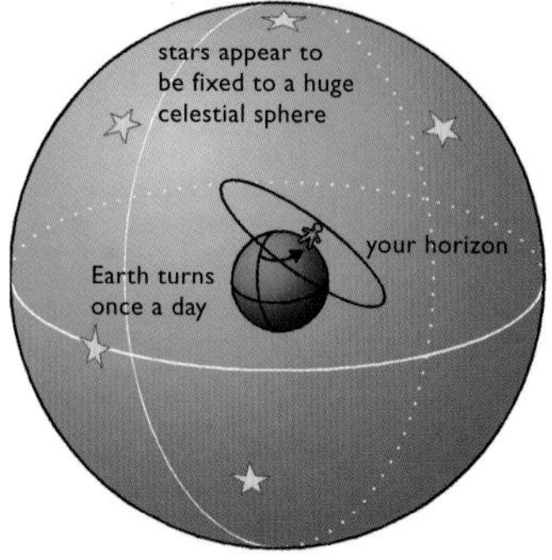

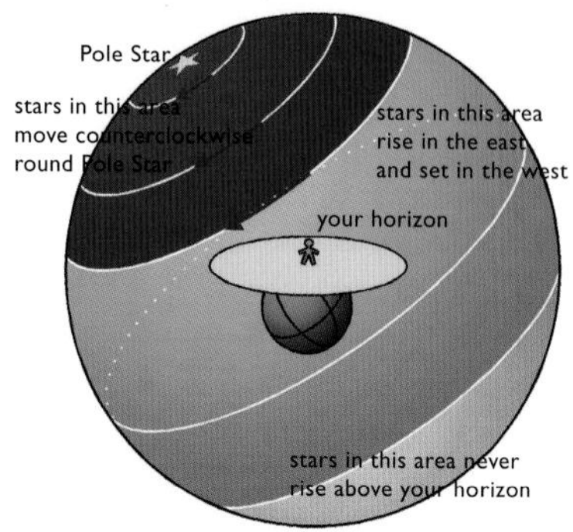

▲ *If you are in mid-northern latitudes, the sky appears to rotate at an angle to your horizon because of the Earth's daily spin. This means that some parts of the sky in the north are always above your horizon, some rise and set daily, and some never appear above the southern horizon.*

close to the north celestial pole, but there is no such help in the southern hemisphere. The bright stars of Ursa Major (better known as the Big Dipper or Plough) and the Southern Cross are convenient signposts to the poles.

The sky rotates counterclockwise around the celestial pole in the northern hemisphere, and clockwise in the southern. Some parts of the sky are therefore visible all night, all year round. Those stars that are currently between the pole and the horizon will be above the pole in 12 hours' time – or, if you like, at the same time of night in six months' time. So as you look toward the pole, you see the same stars whenever you observe. Only the orientation changes. This part of the sky is called circumpolar.

Midway between the celestial poles lies the celestial equator. This is the line that is always above the Earth's equator, and runs in a curve across the sky from due east to due west, reaching its greatest height on the meridian.

Another part of the scene-setting concerns the ecliptic – the path that the Sun takes during the year. You will also always find the Moon and bright planets close to this line. As the seasons change, so does the position in the sky of the ecliptic. On a midsummer night it is low in the sky, the exact opposite of its daytime position when the Sun appears high up. Then at midwinter, when the Sun is low during the day, the Moon and planets ride high. These conditions apply in both hemispheres, even though the months of midsummer and midwinter are different.

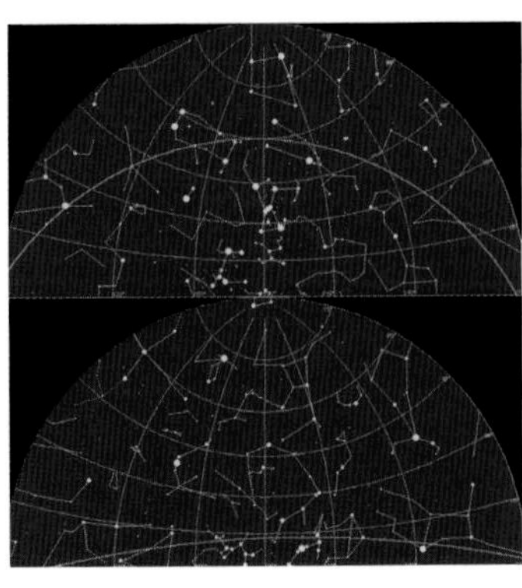

► *The view looking south at midnight in mid-northern latitudes in December (top) and June. The red line marks the ecliptic – the track of the Sun, Moon and planets. In December it is high up, so the Moon and planets are easy to observe, while in June it is low in the sky. Its positions are reversed in the southern hemisphere, or at midday.*

Finally, there is the Milky Way, which occupies another great circle around the sky. Though this band of light is too faint to be seen through the light pollution of many suburban locations these days, the objects within it can be seen with telescopes. The Milky Way is our own galaxy of stars, nebulae and other objects, seen from inside. Think of it as a dense forest, spreading all around us. We are located toward one edge, so in some directions we can just see beyond it to the rest of the landscape beyond. Directly upward, we get a clear view unobscured by trees.

So it is with the Milky Way. In one direction, toward the star patterns of Sagittarius and Scorpius, our view of the center is obscured by gas and dust so we can't actually see the center of the Galaxy, though there are plenty of bright stars in the sky. Most of the stars and other objects

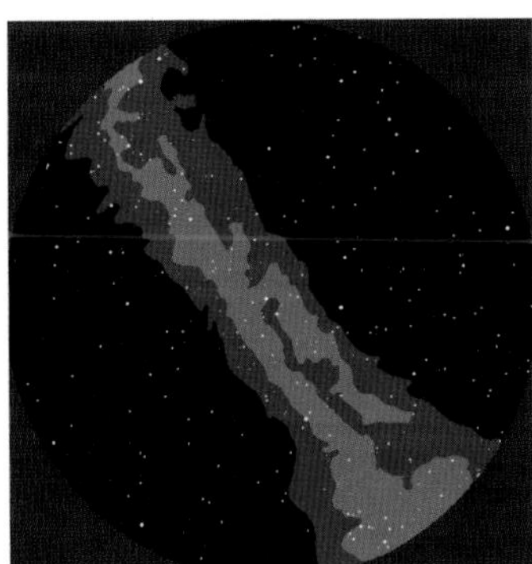

▲ *The Milky Way runs virtually overhead in September evenings in the northern hemisphere, with the rich star clouds of Cygnus and Sagittarius easily visible. In July in the southern hemisphere, the brightest areas are virtually overhead.*

▲ *In April on northern-hemisphere evenings, the Milky Way lies more or less along the northern and eastern horizons and is hardly visible except from very dark sites. A similar situation applies in the southern hemisphere in November.*

we see are comparatively close to us. In the opposite direction, toward Gemini, the Milky Way is much fainter than in Sagittarius. However, the brightest star group of all, Orion, is here, together with its glorious nebula or gas cloud. In the northern hemisphere, Orion is very much a winter spectacle, with the center of the Galaxy visible in summer, while in the southern hemisphere it's the other way around. In spring and autumn, however, our evening skies are much emptier of stars and nebulae as we are looking out of the plane of the Milky Way, equivalent to looking upward in the forest to the clear sky. It so happens that in April and May the stars of Leo and Virgo are on display, and more significantly we get a clear view of many of the distant galaxies beyond our own. In Virgo lies a huge cluster of galaxies, which provide a happy hunting ground for observers of these objects.

All the stars we see in the sky, incidentally, are members of the Milky Way galaxy, most of them in fact comparatively close to us. However, the term "Milky Way" generally refers to the great band of light, consisting of millions of stars too faint to be seen individually.

The constellations

It may seem quaint, in these days of computer-controlled telescopes, that we still refer to star groups that were originally named so long ago that the exact origins of their names are unknown. People have certainly been referring, for example, to particular groups of stars as the Lion and the Scorpion for at least 5000 years, and it is probably safe to say that they got these names in prehistoric times. In the case of these two star groups, or constellations, in particular, it's easy to "join the dots" and see the animal.

One reason why the constellation patterns are still with us is that they provide a very convenient way to know the sky. Just as we might learn our way around a town by establishing a few landmarks and main routes, from which we can fill in the details, so the main constellations give us a framework for finding places of interest. Names mean much more than mere grid numbers, both in the case of city streets and star patterns.

Similarly, in order to find your way around the sky, you need the celestial equivalent of a town plan – a star map.

◀A planisphere is a handy guide to the sky, used by both novice and experienced astronomers. It shows the whole sky visible from a particular place at any chosen time.

Most astronomers have several. One of the most convenient is the planisphere, which is a circular map showing all the stars visible from a particular latitude, with a rotating mask that you can set for the particular date you are observing. Then there are atlases with more detailed maps, some showing the stars visible with the naked eye, while others show stars down to the limit with binoculars or fainter. Finally, there are star mapping programs for computers.

How to use a planisphere

Planispheres can be confusing to first-time users. Before you start, make sure you have one for your particular latitude, north or south of the equator. However, any particular model will work over a range of latitudes, so check with the table below to see which is the most appropriate for your own location.

In principle, the planisphere is easy to use. Match up the date, around the edge of the base map, with the time of your observation (allowing for summer time and your location within your time zone if necessary – see the box "What's the time?" on page 114) and the oval then shows the stars visible at that moment. In theory you should hold the planisphere over your head, though in practice most people will simply hold it in front of them and look up at the sky. However, because the sky appears overhead, the points of the compass around the edge of the planisphere's oval are reversed compared with conventional maps of the ground. The point closest to the edge of the planisphere is due south (or due north, if you are observing in the southern hemisphere). Now you can orient the map according to

► *To use a planisphere, set the map for the time and date then hold it over your head with midnight pointing north. The oval sky area now shows the stars in their correct orientations.*

A RANGE OF PLANISPHERES	
Latitude	**Usable range**
51.5°N	Northern Europe, Northern USA and Canada
42°N	USA, Southern Europe and Northern Japan
32°N	USA, Middle East, North Africa and Southern Japan
23.5°N	Hawaii, Mexico, India, Hong Kong and Taiwan
35°S	South America (South), South Africa, Australia and New Zealand

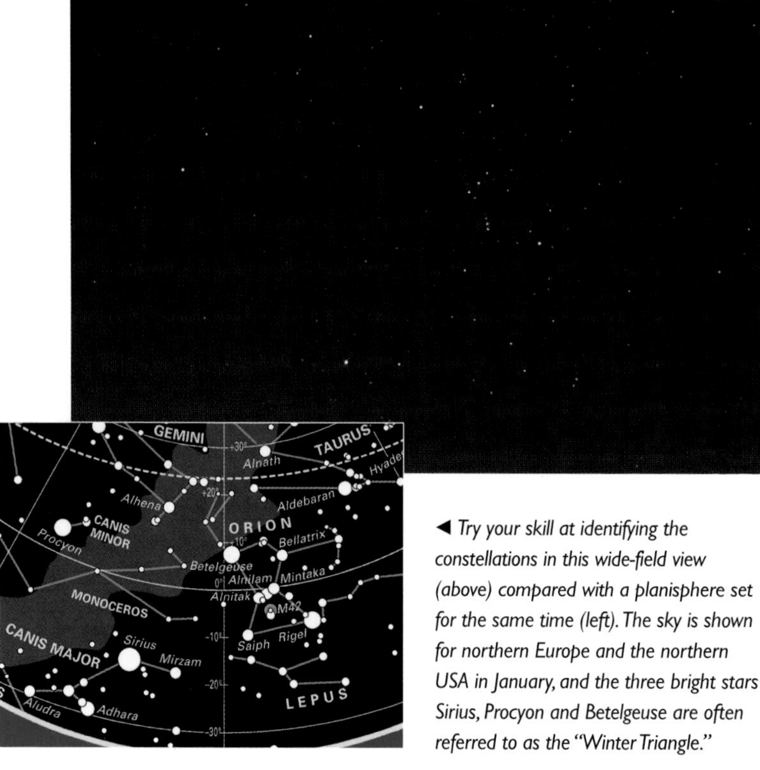

◀ *Try your skill at identifying the constellations in this wide-field view (above) compared with a planisphere set for the same time (left). The sky is shown for northern Europe and the northern USA in January, and the three bright stars Sirius, Procyon and Betelgeuse are often referred to as the "Winter Triangle."*

your location, and the eastern and western horizons should be as shown around the oval.

First-time users may have trouble in relating the scale of the planisphere map with the sky itself. The problem is that the planisphere map is only a matter of centimeters across while the sky stretches from horizon to horizon. However, once you have recognized one constellation, the others should fall into place.

As well as helping you to see which constellations are above the horizon at any particular time, you can also find the approximate position of the Sun, and hence the times of sunset and sunrise. Just place a straight edge between the center of the dial and the date, and the Sun is where that line crosses the ecliptic. Note this position, and move the dial until it touches the eastern horizon. Read off the time against the date in question and you have the time of sunrise. Similarly, when the Sun is on the western horizon, you can read off the time of sunset.

Planispheres usually carry tables listing the approximate positions of the planets Venus, Mars, Jupiter and Saturn month by month for some

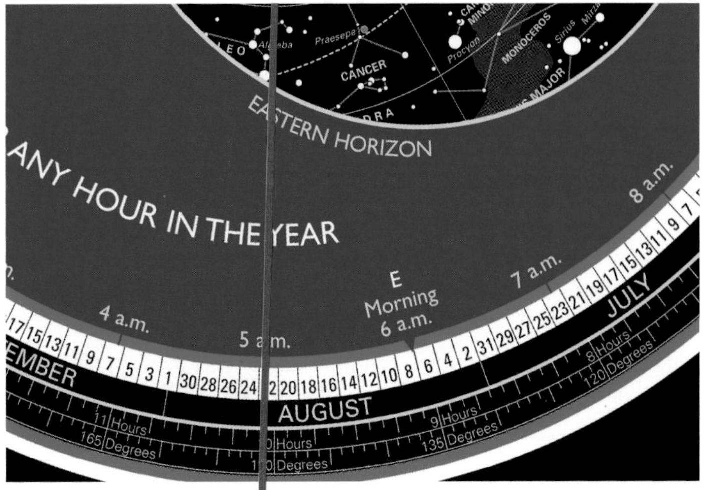

▲ *To find the approximate time of sunrise, lay a straight edge from the date to the center of the planisphere and rotate the dial until the eastern horizon touches the ecliptic at this point. The time shown is that of sunrise – in this case, about 5 am (6 am Summer Time) on August 22 from London, UK.*

years ahead. In each case a position in degrees is given which you can find on the outermost scale, which is in degrees of right ascension – the sky equivalent of longitude (see box "RA and Dec" page 108). These tables will help you identify which planet is which as they move against the starry background.

Star atlases

A planisphere will just show you which constellations are in the sky, and is little help if you want to find interesting objects. For that you need a better star map, such as found in a star atlas. It is probably not a good idea to get too large or detailed an atlas to start with – you need one which shows a considerable area of sky at any one time, so that you can recognize the constellations easily. For beginners there is the *Firefly Night Sky Atlas*, which includes general information on observing and constellation-by-constellation descriptions of objects that you can observe with small telescopes.

With the advent of computer sky programs, you can now produce your own star maps, customized as you want them. The advantage of this is that one system will produce anything from a whole-sky map for your own location at a chosen time, to detailed maps of small regions of sky showing stars as faint as most telescopes will show. Most sky

RA and Dec

Positions in the sky are given by right ascension (RA) and declination (Dec). RA is measured in hours, minutes and seconds eastward from a point in the constellation of Pisces where the ecliptic crosses the celestial equator. It is the sky equivalent of longitude. There are 24 hours making up the complete circle of the sky. Occasionally RA is given in degrees, with 15 degrees to an hour and 360 degrees to the full circle. The reason why RA is measured in units of time is that it is exactly equivalent to the rotation of the sky. So if a star on your meridian has an RA of 3h 30m, in half an hour the star on the meridian will be at RA 4h 00m.

Declination is the sky's equivalent of latitude. It is measured in degrees north and south of the celestial equator, the north celestial pole being +90 degrees and the south celestial pole −90 degrees. These are true degrees of arc, so 10 degrees of declination is the same as 10 degrees of arc. By contrast, an hour of RA is a wide swathe of sky at the celestial equator but just a tiny arc near the poles.

There is one complication. The orientation of the Earth's axis in space changes over a period of 26,000 years, so the whole grid system shifts very slowly across the stars. This movement is known as precession. Star maps have to be redrawn every 50 years or so, to allow for precession, and the date for which star positions are measured should be given whenever accurate positions are needed. All positions in this book refer to the year 2000, which is a standard "epoch" for such measurements. Some older books give positions referring to 1950.

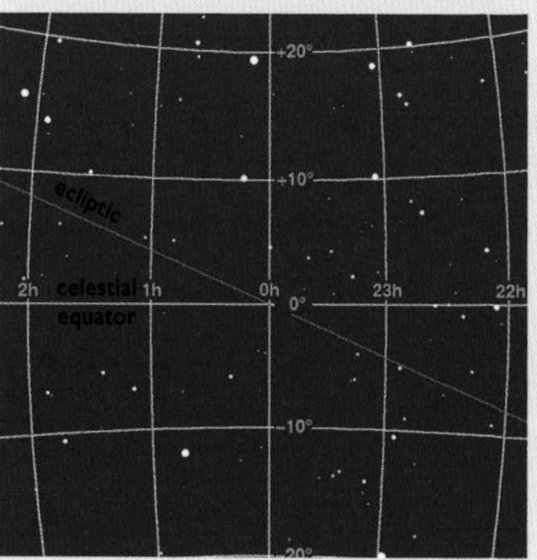

▶ *The sky's coordinates – hours of right ascension, running from west to east, and degrees of declination from north to south. The zero point constantly moves along the ecliptic, and is shown for 2000.0.*

programs include the positions of the planets, and are invaluable for working out what's visible and when, even if you already have a good star atlas.

Armed with all this material, or simply using the star maps on pages 176 to 181, you can now begin to find your way around the sky. For beginners, here's the method that will get you observing within minutes.

Start by choosing the map for the month in question. These maps show the sky at 22:00 local time, not allowing for Summer Time (Daylight Saving Time). If you are observing earlier or later, use the scale of hours along the bottom edge in each case to allow for the time difference. For example, if the date is April 11, choose the map on page 180. The hour of RA corresponding to April 11 is 11h 19m, which is the right ascension of your meridian at 22:00 local time.

Remember that this may be 23:00 by your clock, if you are in the northern hemisphere and Summer Time is in force. If you are observing at 20:45 local time, say, then move 1hr 15m eastward along the RA scale, that is 10h 20m, and this will be the RA on your meridian.

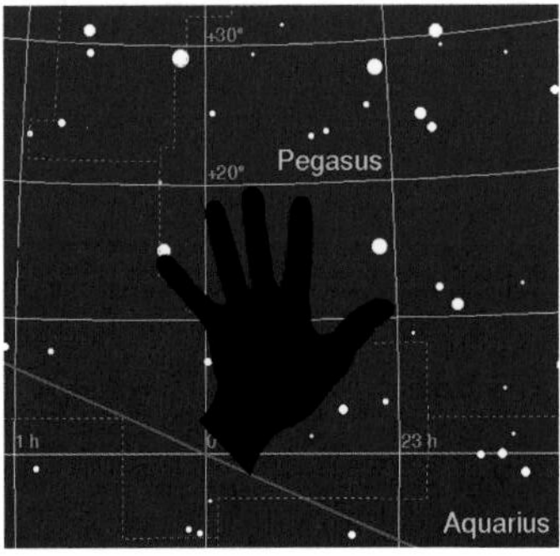

▲ *Get an idea of the scale of the sky by using your outstretched hand at arm's length – here, compared with the map of the Square of Pegasus.*

If you are in the northern hemisphere, hold the book the right way up, and if you are in the southern hemisphere, hold it upside down. The stars in the middle of the map should be somewhere near the middle of the sky looking south in the northern hemisphere, or north in the southern. Their exact location depends on your latitude – in fact, at the equator, they are overhead.

If you are having trouble recognizing the same pattern in the sky as shown on the map, it may be because of the difficulty in relating the scale of the map to the sky. To help sort this out, stretch out your hand at arm's length, so that the thumb and little finger are splayed out. The distance between the tips of the two is around 16–20 degrees. You can check how much this is in sky terms by using the scale of declination along the map edge, which is also in degrees of arc. Another useful arm's-length indicator is the width of a finger, which is roughly a degree.

Finding constellations can actually be far more difficult in a really good country sky than in a light-polluted town because there are so many stars visible. At least in the town your attention is limited to only the brightest. The range of dot sizes which represent stars on maps is always a compromise. The true brightness range of the stars is greater than can easily be shown by symbols: if the faint stars in a good sky are shown by a reasonable size of dot, say 1 mm across, the bright ones would be the diameter of a pencil and would hide a considerable amount of sky. Once you have identified a star group, the scale and brightness representation of the map should become clearer.

Some constellations are easier to recognize than others. Anyone gazing at the evening sky in January or February, for example, can't miss the distinctive shape of Orion. However, a few months later, in April or May, Orion has set by 10 pm, to be replaced in mid sky by the much less obvious Virgo.

The constellations in mid sky, on either side of the celestial equator, progress slowly from east to west, and it is to this area that people direct most of their attention. The ecliptic snakes its way across this zone, so this is where the planets will be found. A full year is needed to view the complete procession of these constellations, observing at the same time in the evening, though you can jump the gun by observing in the small hours of the morning, when constellations are on view that will not be in the evening sky for several months. You can also, of course, get an early view of the planets in this way, before they put on their popular evening shows.

But what of the constellations that surround the pole? These include some of the most familiar star groups, such as the Big Dipper (Plough) in the northern hemisphere and the Southern Cross in the southern. Use the appropriate star maps on pages 176 or 177, depending on your hemisphere, and turn them so that the month of observation is at the top. The map then shows the sky you'll see looking toward the pole – that is, north in the northern hemisphere or south in the southern.

From mid latitudes, where most people live, most of the constellations in your hemisphere's polar map are circumpolar. But the nearer you go to the equator, the smaller the circumpolar area, until at the equator itself the two celestial poles lie due north and due south and there are no circumpolar stars at all.

The only way to really get the hang of the sky and the way it moves is to get out there and start observing. Let's continue the example above for April 11 at 20:45. Choose the map on page 180, and subtract 1h 15m from the RA along the middle because we are observing

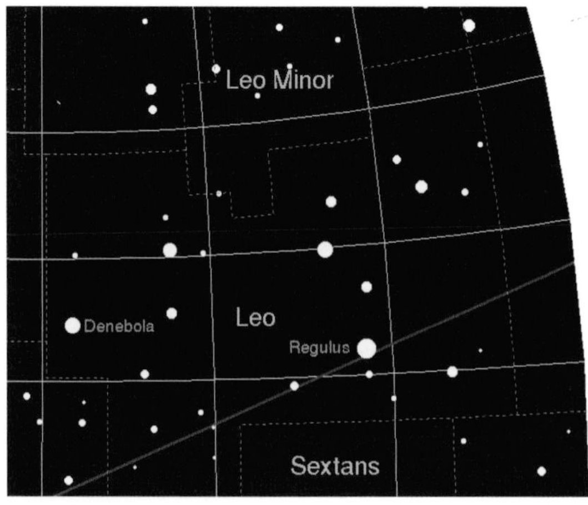

◀ *The main stars of Leo, the Lion. With a little imagination you can see the figure of the animal. The red line is the ecliptic – the path of the Sun. The Moon and planets are found close to the ecliptic.*

Names and numbers

The story of how stars got their names is a long one, and there are whole books on the subject. Basically, most derive from Arabic, Greek or sometimes Latin names that refer to the mythological figure. Denebola, for example, refers to the tail of the lion. There are two other prominent Deneb stars – Deneb in Cygnus is the tail of the swan, while Deneb Kaitos in Cetus is the tail of the whale.

While astronomers often use the old names for the bright stars, the fainter stars either do not have names or they are rarely used. Instead, astronomers refer to star designations set up in the 16th and 17th century. In the Bayer system, the stars in each constellation are given Greek letters, usually starting with alpha for the brightest. The scale then descends to fainter and fainter stars, though there are occasional anomalies. The constellation name takes its Latin genitive form, so Regulus is Alpha Leonis.

In addition to the Bayer letters many stars also have Flamsteed numbers, which are now used only in cases where there is no Bayer letter to refer to. Flamsteed numbers cover stars down to about magnitude 6. Stars fainter than that may be referred to by a simple catalog number, of which the most common is the Smithsonian Astrophysical Observatory (SAO) catalog which goes down to about magnitude 9.5.

earlier in the evening. The map shows a prominent group of stars on the meridian, in the northerly half of the map. These are the stars of Leo, the Lion – a pattern that has been recognized as a lion by people back to the dawn of history, as revealed by Sumerian clay tablets. It is possible that the Sphinx in Egypt is a representation of Leo.

From the northern hemisphere, looking south at Leo, it is not hard to see why. There's a curve of stars that represent the lion's mane as he faces westward, with his body stretched out behind him to the east. Country folk also saw this as a Sickle, while to some it is a backward question mark. For southern hemisphere observers, however, the appearance is the other way round and the pattern is more difficult to identify with a lion.

There is one fairly bright star, known as Regulus, at the base of the Sickle. Astronomers have a rather peculiar way of measuring star brightnesses, which dates back to Greek times. The Greeks called Regulus a star of the first magnitude, and the faintest ones they could see were called sixth magnitude. This system is akin to the way we grade many other things, such as sports leagues and position in a class, but it doesn't tally with modern scientific measurement systems – you would expect greater brightnesses to have higher numbers.

The scale now does have a mathematical basis, in which very bright objects actually have negative magnitudes (see box on page 115). Regulus is now magnitude 1.36, although you will find small variations from catalog to catalog. The second brightest star in Leo, at the opposite end of the lion, is called Denebola and has a magnitude of

2.14, while the third brightest, Algieba, at the base of the lion's mane, is magnitude 2.28.

Having picked out the shape of Leo, you can now use it to find other nearby constellations. In order to find stars to the west of Leo you will have to go to the next map. I have presented the maps in the order they appear in the sky, which may seem logical, except that star atlases usually show the maps in order of rising time. With the maps here, as time progresses you should refer to maps earlier in the sequence, as shown by the months at the top and bottom.

To the upper right of Leo are two first-magnitude stars only about 10 degrees apart. These are Castor and Pollux, which are sometimes called the Heavenly Twins (though usually only in books like this one). They are the bright stars of Gemini, the Twins, which are two ragged strings of stars extending westward from Castor and Pollux.

Between Leo and Gemini lies Cancer, the Crab, but if your skies are light polluted you may well miss it altogether. If it were not for the fact that the ancients wanted to divide the ecliptic into equal portions, for astrological purposes, this area would be known only to astronomers.

By now, the patterns in the sky are becoming evident and anyone with a telescope will be getting anxious to use it on something. Unless

▼ *The bright star is Vega in the constellation of Lyra (see the map on page 179). To its east and north is the easily split double star Epsilon Lyrae. As the inset shows, each star is a double in its own right.*

you have a computer-controlled instrument, you may have problems with finding any but the brightest objects. Experienced observers blithely refer to the method of star-hopping, which means starting from one star which you can find easily, then moving from star to star until you find the object you want. But principle and practice are often different, at least for beginners, who find it hard to be sure exactly where they are looking.

The problem arises through lack of experience in recognizing the difference between the finder's inverted field of view and that through the main telescope. It's one thing to find a bright object such as a planet or even a bright star, and quite another to find fainter stars and then wander off into unfamiliar territory.

The trick here is to practise finding stars of different brightness and then compare the view through the finder and the main instrument. Before you begin, choose the best eyepiece for star-hopping. Ideally it should have a wide field of view, so it is usually the longest focal length you have available. Many commercial telescopes these days provide a 25 mm or 26 mm Plössl eyepiece as standard, which is quite adequate.

Practise on a variety of stars. The most useful are those with a fainter star close by as seen in the finder. When you look through the main eyepiece, notice two things: how bright the two stars appear, and how far apart they are. In cases where the finder shows another star some distance away from the bright one, so it isn't in the main eyepiece field of view, try finding it using the main eyepiece only. Check whether you are successful by looking through the finder.

The star Vega is a good one to practise on, as the famous double star Epsilon Lyrae is 1.7 degrees away, outside the field of view of most telescope eyepieces. You will know when you've found it because it is a widely separated pair of stars of equal brightness, each of which is a close pair in its own right – the so-called Double-Double. Although Vega and the Double-Double are a one-off, there are plenty of other stars to practise on. It may seem tedious to do this, but the fact is that many beginners do complain how difficult it is to find objects.

The other problem faced by anyone in a light-polluted area is finding any stars at all! In this case it may be necessary to star-hop by using the finder – start with a bright star that you can locate in the finder, then move from fainter star to fainter star using the finder alone. You may have to begin your process of familiarization using the difference between the naked-eye view, the appearance of the stars on the star chart and that of the finder.

When star-hopping, the closer you can get to your quarry to start with, the easier your task will be. You may well find it helpful to get a zero-power finder, such as a Telrad or red-dot pointer (see page 157). This will help you to point the telescope at the right part of the sky in the first place, but you will still need experience with recognizing the area of sky using the finder and main eyepiece.

By now you should be ready to begin exploring the heavens for yourself by choosing interesting-sounding objects from the lists in this book and elsewhere. Begin with bright and easily found targets, and only progress to the more difficult objects when you have some experience. In general, star clusters are the easiest deep-sky objects, followed by nebulae (gas clouds). Some nebulae are large and faint, while planetary nebulae (see page 140) are often comparatively bright but quite small, requiring magnifications of over 100 to show them at all. Galaxies are among the most difficult objects, often being both small and faint. The lists in this book give an idea of how easy the objects are to find.

What's the time?

For some astronomical purposes you need to know your local time – which is probably not the time on your watch. Local time is time measured by the Sun as related to your own particular location. In summer, many places operate Summer Time or Daylight Saving Time, which is usually one hour ahead compared with standard time. The thing to remember is that when Summer Time is in force, noon (when the Sun is near the meridian) occurs an hour later than in winter. I say "near" the meridian because the Earth's eccentric orbit around the Sun results in the Sun being up to 15 minutes or so late or early compared with clock time – but that need not concern us here. As far as getting your bearings is concerned, the direction of the Sun at noon is near enough.

To add to the complications of time, you have to know something about your time zone. People living near London have no problems with this, because the world's time is based, by international agreement, on the time at Greenwich, London. This is the zero-point for the scale of longitude around the Earth. The farther west you are from Greenwich, the later the Sun reaches the meridian, while the farther east, the earlier it is. It takes 1 hour to move 15 degrees westward. So at the westernmost edge of Britain, which is about 7 degrees west of Greenwich, the Sun reaches noon half an hour later than at Greenwich, yet the time by everyone's watches (called civil time) is the same as at Greenwich. In Dublin, which is in the same time zone as London, true noon in summer does not take place until around 1.30 pm.

Elsewhere in the world, to find your local time, check how your time compares with that at Greenwich (the international dialing section of your phone book may tell you the time difference). Also work out how many multiples of 15 degrees you are east or west of Greenwich, which will tell you how your time zone compares with the solar time. This becomes important within mainland Europe. In some places in the west of France and Spain, local noon may happen as late as 2.30 pm in summer, and at night the sky view shown on a planisphere will be equally out of step with your watch. In the US, the zone of Eastern Standard Time is also quite wide, taking in Boston to the east and Detroit to the west, not to mention other more extreme locations.

A problem of some magnitude

Mathematically, a brightness range of 100 is five magnitudes. This means that each magnitude is 2.512 times brighter than the next (2.512 is the fifth root of 100). The scale has the advantage that it can deal with a huge range in actual brightnesses using manageable numbers.

It may seem odd that there is no accepted reference point, just as the freezing and boiling points of water define the temperature scale. But there is no star that is always taken to be exactly magnitude 0.00, for example, because stars may vary slightly in brightness. When the magnitude scale was first set up scientifically, it was made to fit in with the generally accepted existing brightnesses, which led to some objects having negative magnitudes. Each new catalog of star brightnesses is adjusted to fit in with the established values – in effect, the whole scale sets its own reference point.

Object	Visual magnitude
Sun	−26.78
Venus at brightest	−4.7
Brightest star (Sirius)	−1.44
Vega	0.03
Spica	0.98
Castor	1.58
Polaris	2.02
Naked-eye limit	6
Typical binocular limit	9
Typical visual amateur telescope limit	14
Typical amateur CCD limit	20
Typical giant telescope limit	30

It is quite fitting that the entire range of brightnesses that we have to deal with in astronomy extends more or less evenly on either side of magnitude zero. Between the Sun and the faintest objects visible with giant telescopes is a brightness range of about 100 million billion billion.

The best way to get to grips with the magnitude scale is to have some idea of the magnitudes of various objects in the sky, as shown in the table. Although the original scale was worked out on the basis that the faintest star visible with the naked eye is magnitude 6, some people can see stars fainter than this, particularly in very clear desert skies.

6 • WHAT TO OBSERVE, AND HOW

With your telescope you can see with your own eyes a good selection of the objects in the Universe. Broadly speaking they divide into Solar System objects and deep-sky objects. The Solar System is the Sun's family of planets, which are, in astronomical terms, in our own backyard. Space probes have visited some of the planets, and there is no technical reason why people should not go there within the next century or so.

Beyond the Solar System, however, there is an enormous divide before we get to the deep-space objects. Even the closest of these are so remote that they are beyond easy reach, and most are so distant that not even science fiction technology, as employed in Star Trek for example, is adequate for visiting them.

But you can explore many of them with your telescope, in some way. In the case of nearby objects, such as the Moon, you can get so close that you can imagine yourself flying in a spacecraft over its surface. But remote objects, such as galaxies, are so distant that all any telescope will show is a faint, fuzzy patch of light that you are happy simply to have located at all.

▼ *The first-quarter (left) and last-quarter (right) Moon, with the principal seas and a selection of features named. Use these few landmarks to find your way around at first, then fill in the details with a larger-scale map.*

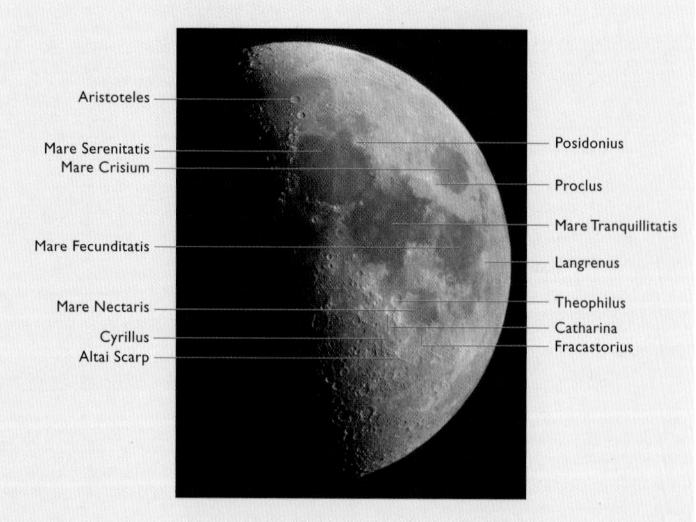

Ideally, I would like to give a complete outline of each object – what it is, why it looks like it does, how big it is, and so on. But that is outside the scope of this book, so if you want to learn more about the objects themselves please refer to a more general astronomy book or go to the website addresses provided.

Our tour of the Solar System begins with the closest object, the Moon.

A world of mountains and craters

There are very few objects visible through a telescope that are virtually guaranteed to look amazing. The Moon is top of the list, and it is available for viewing at some point in its orbit around the Earth for about three weeks in every month. It is a shame, in some ways, that to most astronomers the Moon is a wretched nuisance, or at least an opportunity to get a good night's sleep safe in the knowledge that its light is drowning out views of all other objects. The life of both amateur and professional observers is governed by the orbit, every 29.5 days, around the Earth of our only natural satellite. And there is a lot to be said for lunar observing – it acts as an excellent training ground for the more demanding planets, which is why I have devoted a lot of space to it.

http://www.nineplanets.org/luna.html

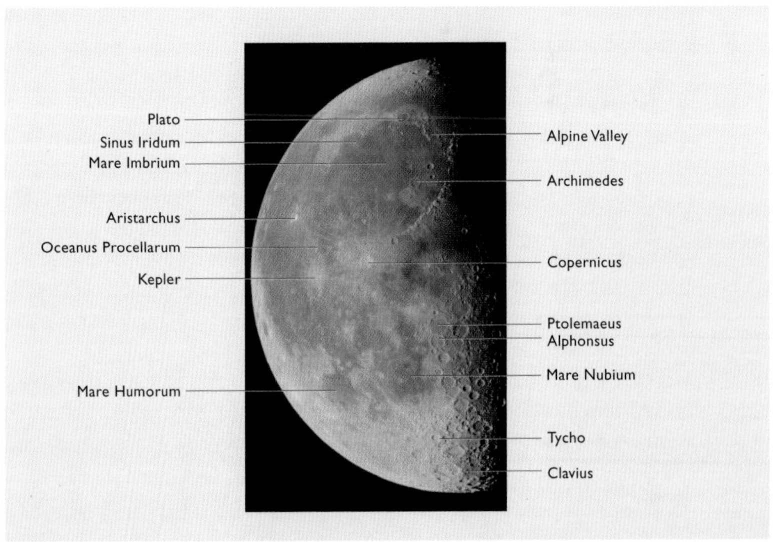

To the lunar observer, the phase of the Moon is crucial. The line between darkness and shadow, the terminator, is the most interesting part of the Moon at any time. Its craters and mountains throw long shadows at this point, making them highly visible. The next night, the terminator has marched eastward, and the features that were so obvious last night are scarcely distinguishable against the mottled background.

The features of the Moon have a romance all of their own. The first observers to use telescopes gave them fanciful names, not knowing what they really were, so we get – in Latin – seas, lakes, mountains, marshes and bays, plus a dignified gathering of classical astronomers whose names have been given to the craters. In reality, of course, the dark seas are lava flows of basalt that solidified billions of years ago having flooded most of the Moon's lowland terrain. The craters are the result of millions of impacts of bodies, large and small, on the surface soon after the formation of the Solar System.

No photograph can do justice to the sight of the Moon on a good night through even an inexpensive telescope with good optics. The blackness of the shadows is total, while the peaks shine out brilliantly. Unlike most other astronomical objects, where details are subtle and pale or at the extreme limit of visual acuity, the Moon really hits you squarely between the eyes.

What you can see is a fascinating landscape. If you begin your observing when the Moon is a crescent you will see one prominent sea, the oval Mare Crisium (Mare is pronounced "mah-ray"; the plural of mare is maria, pronounced "mah-ree-ah"). As the month continues, more seas come into view. One of the most notable, the Mare Imbrium, starts to become visible in the northern half of the Moon just after first quarter, when the Moon is at half phase (that is, a quarter of the way round its orbit). The entire eastern half of the Moon (eastern as seen in the sky) is dominated by the Oceanus Procellarum, a vast lava plain. Get to know a few seas and the others fall into place.

They also act as convenient guides to the craters, of which hundreds are named. Only the lunar experts bother to remember the finer details of lunar geography, but the main seas, craters and mountains soon become familiar territory.

If the Moon is at its most interesting around the terminator region, Full Moon, by contrast, is bland. One view of Full Moon lasts you quite a long time, partly because it is less exciting than at other times, and

www.fourmilab.ch/earthview/lunarform/lunarform.html

► *The crater Pythagoras, drawn by the author using a home-built 87 mm reflector. You do not need much artistic ability to get quite an acceptable drawing.*

also because you can't see anything else for a while after looking at the Full Moon – it really is bright, though not dangerously so.

After full, the same features can be seen on the terminator, night by night, as before full. But the change of direction of illumination can make them look quite different. In fact, it is rare to see the same feature looking the same twice running, because in addition to the daily shift of the terminator, the Moon itself shifts slightly from night to night – a wobble from east to west, or north to south, called libration.

The Moon always keeps the same face to Earth – what is known as captured rotation. This is because of tidal forces between the Earth and Moon early in their history. But libration means that we do get slightly different views each lunation (cycle of phases).

Having gazed and marveled at the lunar landscape, what can you do next? The Moon's surface is to all intents and purposes unchanging, and is well mapped, so observations are mostly for amusement only, but there is a great satisfaction in recording what you see. And the great news is that drawings of the Moon are quite easy to make and actually have advantages over photographs or CCD images. The same applies, in fact, to the other planets, and the Moon is a good place to get experience in sketching at the telescope because it is so clearly visible.

The reason why drawings are better than photographs lies in the seeing, caused by atmospheric turbulence. The eye is very good at picking out details seen in a fraction of a second, when the seeing steadies unpredictably, whereas a photograph requires an exposure time of a considerable fraction of a second, by which time the seeing

Drawing at the telescope

There is a great tradition of making drawings while observing, and it persists even today. Drawings of planets and deep-sky objects are still posted on computer forums as well as hi-tech CCD images. Some can be works of art, while others are simply records. Even if the drawings are of no scientific value, they can be fun to make and act as a good record of what you saw on a particular occasion.

The apparatus is simple. You need a clip board to provide a solid surface, carrying its own light. Clip-on LED lights are available from astro suppliers to provide just the right level of red light, often variable in brightness, for drawing without destroying your night vision.

Observers generally use pencil, each observer having a personal preference for soft, medium or hard lead. Pencils with an eraser on the end can be useful, and your finger can be helpful when smudging an area of tone to widen it or make it more subtle.

Whatever the object you are drawing, the usual approach is first to get down the overall structure in outline, then add progressively finer details. In the case of the Moon or a planet, just indicate the major light and dark regions, and worry about shading and detail once you have the structure right. If you are anything like me, just getting the overall positions in place is a challenge.

For open star clusters, begin with the brightest stars and then put in the fainter ones. Globular star clusters are more of a problem, as there can be literally hundreds of stars visible. However, the surrounding stars are important when establishing the scale and orientation of your drawing compared with other drawings or photos, so try to get as many real stars as you can before resorting to a general blur.

When drawing planets, you can at least give yourself the outline of the planet before you begin. It is actually quite rare for this to be a circle. Only Mars at opposition is a circle – all the other planets, and Mars at other times, either show some phase or are flattened, like Jupiter.

Saturn is the biggest problem. Not only is the planet noticeably flattened at the poles, but the angle of the ring system changes from year to year. The answer is to prepare a template in advance by either copying the orientations shown in the monthly observer's pages of astronomy magazines, or using a computer drawing program.

The paper you use can be high-quality drawing paper if you wish, though many people simply use ordinary sheets of paper for their sketches at the telescope. It is not unknown for observers to make their sketches on the same observing books, with ruled pages, in which they keep records of all their observations. Observing books are a good idea, as they mean that you keep all your records together rather than scattered on separate sheets.

Whatever you draw, make sure you always add the date, time (not forgetting to specify the time system or zone), object details if not obvious, seeing conditions, instrument and magnification used, and anything else that might be relevant. Also add the orientation of your drawing – particularly in the case of deep-sky objects.

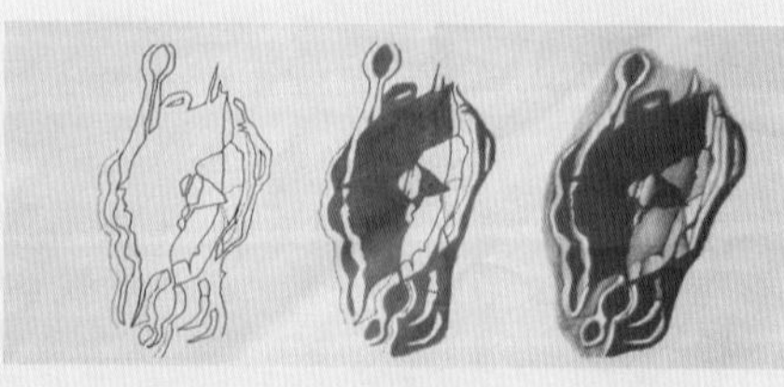

◀ *Stages in drawing the lunar crater Piccolomini, made by Peter Grego using a 60 mm refractor. The same basic principles apply to virtually any astronomical object.*

has blurred the image. Even if you are monitoring the view, the chances of catching a brief moment of good seeing are very small. The only hope is to take large numbers of shots and select only the best. This is feasible using a video camera or webcam (see pages 162–3), and the detail on images made using these can rival or even exceed that on a drawing made with the same instrument. But drawings are cheap and fun to do, so the choice is up to you.

You don't need great artistic skill, though for the more pictorial results it can be worth developing your talents. See the box on drawing opposite for general details. Sketch in the outlines of the major features first, including the positions of the shadows, then concentrate on the finer details. You will need to add shading at some stage if you want a pretty result, but the most important thing when at the telescope is to concentrate on recording what you see rather than achieving a perfect artwork. It is perfectly acceptable to produce a finished result afterward, as long as you are sure of your interpretation of your sketch.

You will have to work fairly quickly, because you will be surprised by how much the shadows can move over even half an hour. A crater that was in darkness when you began your sketch is already catching the first rays of the Sun as you are starting to fill in the details. This is one of the delights of lunar observing – watching the slow revelation of new features. But for the budding artist, it is a good idea to concentrate on a small area only to begin with.

Venus and Mercury – fugitive features

If the Moon is easy to observe, Venus and Mercury present the observer with major challenges. Although Venus is unmistakable when it's in the evening sky, when it is often called the evening star, through a telescope it is a secretive planet as far as detail is concerned. Mercury, the closest planet to the Sun, is even more private. Many amateur astronomers admit that they have never even seen it, let alone observed it, because of its habit of skulking down on the horizon.

Both planets are closer to the Sun than we are, which accounts for their behavior. They can only ever be seen on one side or the other of the Sun, and usually in a twilight sky. Imagine that you could see them in the sky at midday, when the Sun is high up. Their orbits are seen roughly edge-on, so they shuttle from one side of the Sun to the other. In fact, we can only observe them when they are at least 20 degrees or so from the Sun, which in the case of Mercury means at the extremes of its orbit. When it is on the west side of the Sun it is said to be at western elongation – although this means that it will be visible in the eastern sky, just before sunrise. To see it in the evening

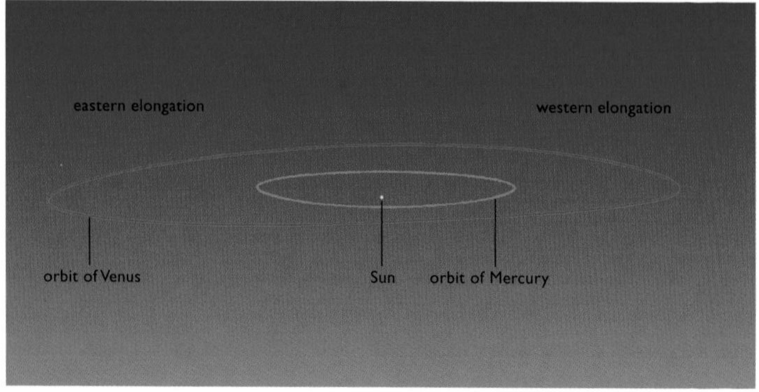

▲ *Typical orbits of Mercury and Venus around the Sun, drawn to scale, as they would be seen in the sky. The rather skewed appearance is caused by the Earth's own movement around the Sun.*

sky it must be within a few days of eastern elongation. Mercury is only visible for a week or so at a time, as a bright star near the horizon in twilight, and is never visible in a dark sky.

Venus, however, is another matter because it can travel nearly 50 degrees from the Sun. It can be seen in a dark sky, when it is near a maximum elongation, although it's often low down under these circumstances.

So the way to observe Mercury and Venus is to do so during daylight – even when the Sun is high up. Venus in particular is easy to see through a telescope during daylight because it is so bright, but unless you have crystal-clear skies, Mercury will be more of a challenge even at its maximum elongation.

Picking up the planet is easy enough if you have a telescope equipped with setting circles (see page 97). Even Go To telescopes are handicapped, because they require reference stars for their alignment, which of course are largely invisible by day. If you use a fixed observing location for your Go To telescope you can set up a nearby marker due north (or south) on which you can initially align and level the telescope. Program the date, time and location accurately and make sure the mount is level, then simply accept the telescope's own choice of its reference star positions. It should then find Venus for you, at least within the field of view of the finder. But you must have left the telescope focused on a celestial object with a low-power eyepiece beforehand, because if the telescope is defocused you may not be able to find even Venus. Focusing on nearby objects is not the same as focusing on infinity.

Many Go To telescopes allow you to "park" the telescope after use, which means that they will remember their alignment settings. However, this means leaving the telescope in position overnight with power on, which may not be convenient.

What can be seen on Mercury and Venus? Both planets go through the same range of phases as the Moon, though neither can be seen at the equivalent of full, because they are then on the far side of the Sun. The closeness of Mercury to the Sun means that it is usually only visible when it is near half phase. Venus, however, can be seen as a thin crescent for a short period as it nears the Sun at the end of its evening-appearance season or the beginning of its morning appearances.

▲ *Mercury is only ever visible close to the horizon at twilight. This photo was taken in spring from the UK, which is when the ecliptic is steeply inclined to the horizon from the northern hemisphere.*

▲ *An amateur CCD image of Mercury, taken by Maurice Gavin using a 300 mm reflector. You can see the phase but no other details.*

▲ *The crescent phase of Venus, drawn by David Graham using a 150 mm reflector with a power of 166. These markings are about the most you will ever see on Venus.*

With Mercury, the phase is just about all you will ever see, and even then it can be hard to make out. There are markings to be seen on Mercury, because it is a cratered world resembling the Moon, though without the familiar seas. You need a good telescope and perfect seeing conditions, and if you do see anything definite you will be very fortunate.

Despite its brilliance, Venus is little more rewarding at first sight. It can be beautiful seen against a deep-blue sky, but you have to be persistent to see any variations in its countenance. But Venus has its share of oddities. The reason it is so bright is that it is completely cloud-covered. There are faint streaks in the clouds, which amateurs have recorded from time to time.

Changes in the shape of the terminator are easier to spot. Occasionally it appears slightly crooked or bumpy, and around the time of dichotomy, when Venus should be completely bisected by the terminator, which should be a straight line, it shows a smaller phase than predicted. Presumably these effects are the result of fine structure in the cloud tops that we cannot resolve. Another curiosity is the Ashen Light, when the dark part of Venus appears lighter than the sky background. Is this just an optical illusion? The jury is still out. Occasionally, when Venus is very close to the Sun, the crescent can extend all the way around the disk, presumably because of light that is refracted by its dense atmosphere.

If you are taken by Venus, you will be in select company. It is very much a planet for drawing. Webcam images are worth attempting, particularly using infrared filters.

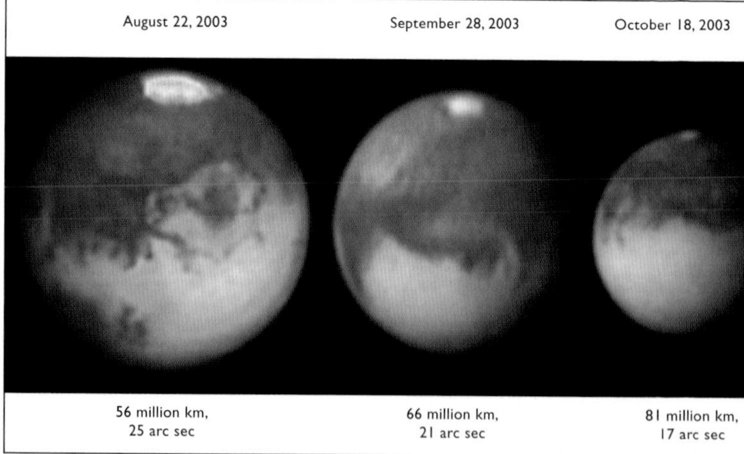

Mars – worth the wait

Observers of Mars have to be patient: because it is the next planet out from the Sun from Earth, it plays a continual game of tag with us. All the other planets are visible every year, one way or another, but Mars manages to give us the slip and can only be seen well every two years.

Mars is known as the Red Planet, which leads people to expect a stronger color than they actually see. In fact, it has a pale salmon-pink disk that at first glimpse, even with a reasonably large telescope, appears rather bland. The sad truth is that illustrations often emphasize the color and contrast of Mars' features. But take a closer look and, if the seeing is good, you should see the glint of a white polar cap, and dusky markings on the body of the planet.

Old books used to insist that these markings are green, but speaking personally I have never seen them as anything other than a darker shade of salmon pink. The supposed green color gave rise to a whole mythology about Mars, that the dark markings, which change in appearance with the Martian seasons, could be evidence for lowly life forms, similar to the blue-green algae that are believed to be among the most primitive life forms on Earth. Today, spacecraft images show that the changes in the markings are the result of wind-blown dust covering and uncovering darker rocks, and the hunt for life has shifted below the planet's surface where there are believed to be layers of permafrost.

▼ A sequence of webcam images of Mars made during its very close opposition of 2003, showing the change in size of the planet during the year.

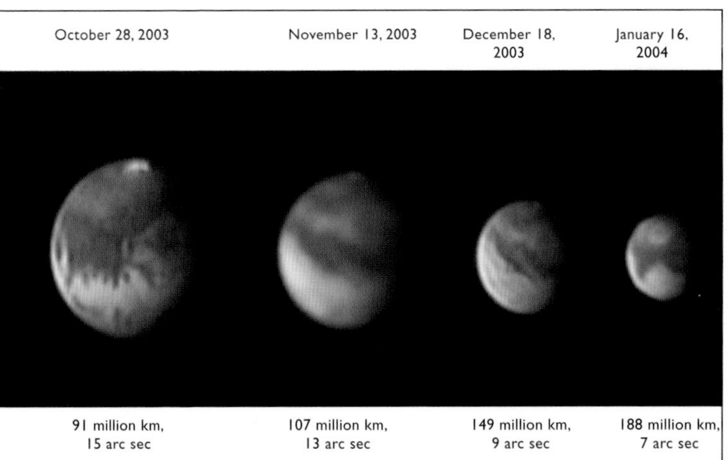

The telescope view of Mars shows that the dark markings are concentrated in the planet's southern hemisphere, which appears at the top when seen in an astronomical telescope in the northern hemisphere of Earth. It is great fun to sketch the markings you see and to try to tally them up with a map of the planet, or most realistically one of the Hubble Space Telescope photos. Map projections tend to give greater prominence to the poles, which are much more foreshortened when you see them from Earth, whereas the HST views are more representative of the telescopic view. If you study the planet carefully, under good seeing conditions, you can actually get a view that is not much worse than the HST images, and webcams now also give impressive results.

The planet rotates in just 24 hours 37 minutes. New features come into view every hour or so, but the only way to see the whole planet is to be patient. The following night at the same time you will see virtually the same side, but to see the opposite face you must let those 37 minutes accumulate – about three weeks should do it.

Asteroids

We hear a lot these days about the asteroids or minor planets, particularly those that come close to the Earth instead of orbiting between Mars and Jupiter like most of the others. At one time they were a very minor field of interest for amateurs, but now they are in fashion, for several reasons. One is that the use of CCDs (see page 171) has made it possible for advanced amateurs actually to find previously undiscovered asteroids with comparative ease. Another is that computer sky programs can provide finder charts for asteroids for any night.

If you have such a program, you should be able to locate the field of view using the general techniques of star-hopping. Many asteroids are quite bright – Vesta, the brightest, is even just visible to the naked eye on good nights – so the problem is not finding them but knowing when you have seen them. Your sky program should give not only the asteroid's position but also its magnitude. If a star of the right brightness is unmistakably visible in exactly the right spot you have found your asteroid, but even so it would be a good idea to make a sketch of the area and observe again a night or two later. The extra star should have moved. Most asteroids appear starlike, though when it is at opposition the largest, Ceres, can display a tiny disk that can just be recorded using a webcam or CCD, though not even HST images show much detail on it.

http://www.cfa.harvard.edu/iau/mpc.html

▲ *This is how Jupiter and its bright satellites appear through a small telescope. Sometimes the orbits of the satellites are in the same plane, as here, and at others their orbits are more inclined.*

If you want to discover your own, and have the honor of naming it yourself, you will have to make regular and deep images of the ecliptic, look for additional objects that move in the right way, weed out all the objects that are already known by referring to online databases, measure the positions precisely and painstakingly using the correct specialized software, then reobserve the object and repeat the positional measurements over a number of years. If you can go through all that, you deserve the right to name it, subject to approval by the International Astronomical Union.

Giant Jupiter

If Venus and Mercury are bland and Mars is usually tiny, Jupiter makes up for them all. It is in the sky every year, has the largest disk (apart from Venus as a thin crescent) and shows a wealth of detail that is forever changing. Even the tiniest astronomical telescope should show something of interest on Jupiter.

As the largest planet in the Solar System, Jupiter's size is obvious as soon as you look at it. Despite being ten times farther away than Mars even when both planets are at their closest, Jupiter appears twice the size. The markings, which mainly consist of dark belts and light zones, are much more distinct than those of Mars. And to add to the spectacle, Jupiter's four largest moons continually shuttle back and forth, providing a constantly changing view.

▲ *Jupiter, imaged with a cheap Philips ToUcam webcam by Dave Tyler using a 200 mm reflector made in the 19th century. The image is a combination of the best 800 frames out of 1200 recorded in 2 minutes.*

Jupiter shines with a bright creamy color that is quite distinctive. Actually, each of the planets has a characteristic appearance that sets it apart from the others and from the stars, with the exception of Mercury, which could be mistaken for a star. As well as their color and brightness, planets rarely twinkle. This is because they have disks, so their light is less easily disturbed by atmospheric turbulence than that of stars, which are points of light.

Even in the finder, Jupiter looks different, with its obvious disk and retinue of bright satellites. In the telescope, just a few moments' study shows detail. With an aperture of 75 mm or larger you can easily see spots, both light and dark, within its bands and zones. The most famous, the Great Red Spot, is a great oval storm.

Sometimes the width and intensity of the belts and zones change, for reasons as yet unknown. New spots may develop. The Great Red Spot may become dim and hardly visible, or it may become very prominent, so don't always expect it to be there.

One of the most dramatic events ever seen in the Solar System took place in July 1994 when pieces of a comet named Shoemaker-Levy 9 hit Jupiter over a period of many days. The resulting dark stains at the impact sites were clearly visible to amateur astronomers the world over. No similar event has ever taken place during the time astronomers have been observing the planets, though various dark spots seen in the past may have had a similar cause. So there is just a chance that you may be the first to witness some strange new event. The Solar System can be a dynamic place.

Superb Saturn

Everyone wants to see Saturn. The amazing rings surrounding the planet are unique in the Solar System, though the other giant planets, Jupiter, Uranus and Neptune, all have rings that are invisible to amateur observers.

As a result, Saturn is a planet of beauty. The rings can be seen with small telescopes, and even high-powered binoculars show that the planet is not a simple disk, though you need a power of about 50 to show their true nature.

If it were not for the rings, Saturn would be a yellowish planet slowly plodding round the ecliptic in 29½ years, and therefore remaining in the same general area of the sky for several years at a time. It is as bright as all but the brightest stars, so it stands out among the stars as seen with the naked eye. Its disk varies between 15 and 20 arc seconds across, but the rings make it up to a respectable 45 or so arc seconds at opposition.

Despite its instant appeal, Saturn is not in the same league as Jupiter or Mars when it comes to variations. Occasional changes take place on

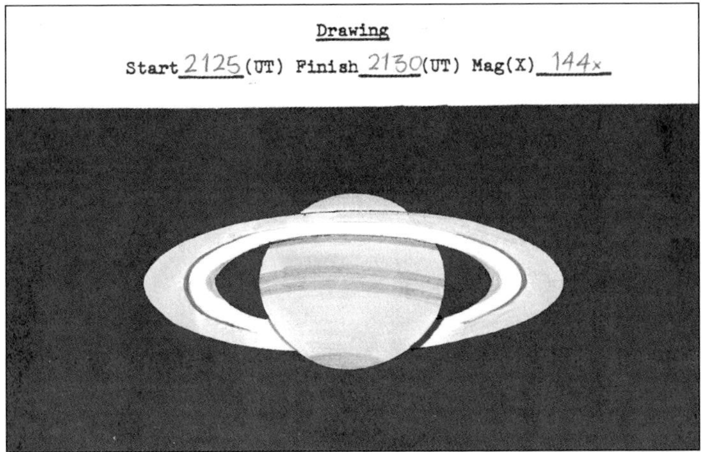

▲ *David Lloyd made this drawing of Saturn in pencil on a photocopied blank of the disk and ring orientation. He used a 150 mm reflector and a power of 144.*

the disk, such as the appearance of white spots, but they are neither as frequent nor as easy to see as those of Jupiter.

The rings remain pretty well constant in appearance, except that they change their angle to the Earth and Sun throughout the 29½-year period. Saturn has five satellites that are fairly easy to see in amateur telescopes. The largest, Titan, is a fascinating world in its own right, and was the target of the Huygens space probe which landed there in 2005. At eighth magnitude, Titan is visible with most telescopes. A fainter satellite, Iapetus, is curious because one side is much darker than the other, though no one is certain why. As a result, it's always brighter when it's on the western side of the planet than when it's on the east. It takes just under three months for every orbit of Saturn, so in some months when it is magnitude 10 you should be able to glimpse it with a 50 mm telescope, while in others when it is magnitude 12 you will need at least a 75 mm telescope.

The outer worlds

Uranus and Neptune are bright enough, at magnitudes 6 and 8 respectively, to be found and observed with small telescopes. Both planets appear bluish, even in small telescopes. Uranus displays a tiny bland disk, but Neptune is more difficult, with a diameter of only about 2.3 arc seconds. Observing these planets is more a matter of being able to say you have done it than seeing something spectacular, but that's no reason for not trying.

So it is with remote Pluto, though here the observation is much more of a challenge because it is about 14th magnitude. Even with a 250 mm telescope it is near the limit of visibility. It appears completely starlike, and all you can do is to pick out which object it is, just as with asteroids. Find it as you would an asteroid, by using a computer sky-mapping program to plot a map of the area, then observing on successive nights to check that your suspicions as to which object it is are correct.

Comets

Until the appearance of Comet Hale-Bopp in 1996–7, few people had seen a comet. It was a beautiful sight for several weeks and almost everybody in the northern hemisphere who wanted to could see it, even from city skies. The story of its discovery, by two amateur astronomers independently, added to its appeal.

Such bright comets are rare and unpredictable events, but there are usually a few much fainter comets to be seen at any time. Most are not easily visible from light-polluted skies, and you need to have access to up-to-date information to know where to look. The popular astronomy magazines give positions of the brightest ones, but for the fainter run-of-the-mill comets you need to get predictions month by month, which these days means using a specialist website.

You don't need a telescope at all for good views of bright, spectacular comets. Instead, use binoculars to see the whole comet. But for regular comet observing, a telescope is essential. Those devoted souls who aim to find new comets may opt for wide-field, fast reflectors that will cover a large area of sky at one glance, or very large binoculars such as 25 × 105s, but for the rest of us, who are content to observe the comets that someone else has found for us, more ordinary telescopes are fine. The important thing is to get good contrast between the comet and the background, so refractors may have the advantage.

First, however, you must find your comet. Traditionally, you plot out the published positions of the comet – known as an ephemeris – on a star map. An ephemeris often gives the position of a comet only every five days, so a certain amount of guesswork and interpolation is needed. You also have to remember that the positions are for 0h Universal Time on the day in question, so if you are observing from a different time zone, and on the evening of that day, the comet will have moved on significantly. In fact, for North American observers you can often take the next day's position for your evening observation, since 0h Universal Time on the 22nd, say, is 19h Eastern Standard Time on the 21st.

http://www.cfa.harvard.edu/iau/cbat.html

Today, it is easier to enter the elements of the comet's orbit, if you have them, on to a computer sky-mapping program, and make a print-out of the area of the comet at the time of your observation for you to take to the telescope. And of course the computer-controlled telescopes should find the object for you, given the correct elements. But finding the position in the sky and actually seeing the comet are not the same. Often, your view will be of a blank bit of sky with maybe a few stars. Comets can be low-contrast objects, and they can be small. Even computerized telescopes do not always find the precise spot, so it can help to have a star map to confirm what you should be seeing, and where the comet should be in relation to the stars. Occasionally, predictions of newly found comets may be slightly in error, so study the whole area carefully.

But unless you have good, dark skies don't be surprised if you don't spot the comet – or to hear that someone else saw it on the same occasion and claimed it was easy to see. Even quite bright comets are

▼ *A print-out from the SkyMap program of positions for Comet C2001 Q4 NEAT, visible near Ursa Minor in 2004. The comet's orbital details, obtained from the IAU website, were entered into the program's comet database.*

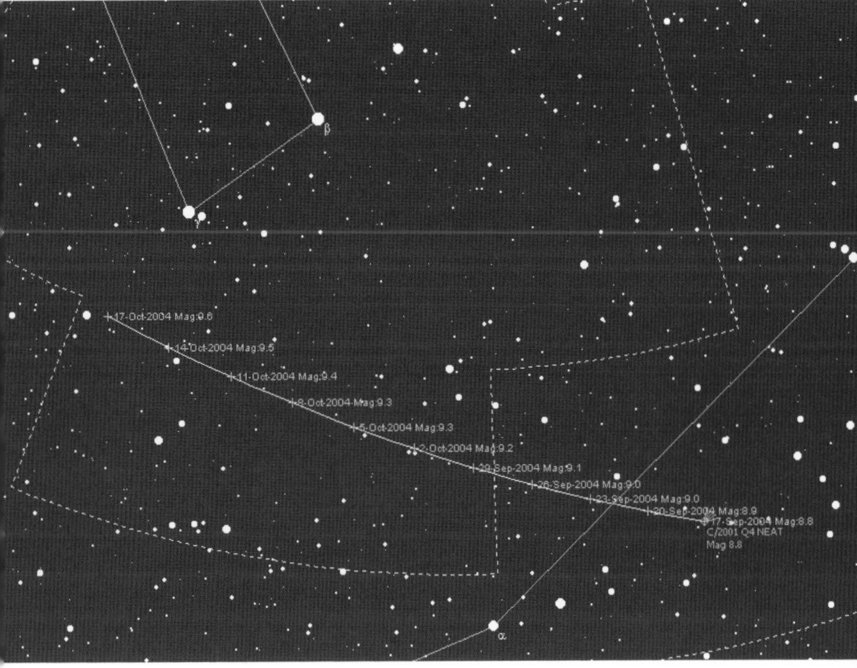

▲ *Comet Hale-Bopp (above left), which was at its brightest in 1997, was the best-seen comet for many years. But comets usually look much more like this 1990 view of Comet Levy (above right).*

notorious for disappearing into the background, and your local sky conditions are all-important.

Do not expect to see a miniature version of Hale-Bopp in the eyepiece. The typical comet is more likely to be a circular, fuzzy blob, possibly with a bright center, and maybe elongated in one direction rather than having a sweeping tail. Some are small and condensed so you need a power of around 75; others are large and diffuse, requiring a low power and a dark sky. Often, comets are only visible in a twilight sky, since they brighten up as they near the Sun, so even though your target may well be within the range of your telescope it may be impossible to see against the twilight, or too low down in your sky to be seen.

Having found your comet, look for signs of activity at its heart – the nucleus. Actually, the starlike point you may see is not the true nucleus, the icy body that produces the gas, but the much larger cloud of gas surrounding it. There may be a tail, or maybe more than one, so look carefully using averted vision.

The Sun

Astronomy is a pretty safe occupation, but where the Sun is concerned there is a real danger that you may either damage your eyesight or at worst blind yourself. Don't take any chances. The safest and simplest way is by projecting its image on to a piece of white card. Before you do this, cover the finder telescope so that there is no danger that someone will inadvertently look through it, or of damaging its eyepiece and crosswires with the heat.

Also, cut the aperture down to 100 mm or less if your telescope is larger than this in the first place. Many instruments have covers with additional caps covering a smaller aperture, but if not, make your own from cardboard. In the case of a reflector or SCT, cut the hole off-center to miss the vanes of the secondary mirror or the central obstruction. It should not matter that the aperture is not central – each part of the mirror contributes equally to the image in a well-made telescope. And although the image brightness suffers when you stop down a telescope, the resolution (the detail you can see) is usually restricted by the seeing, rather than the aperture, in any case.

The reason for stopping down is to reduce the heat that will be passing through the eyepiece. It may be a good idea to use a cheap, uncemented eyepiece (see page 146) if you plan a lot of solar observing by projection. But do not use eyepieces with plastic barrels, which will melt.

Now at last you are ready to project the image. Fit a low-power eyepiece and turn the telescope sunward. You can't use the finder, but look at the shadow of the telescope or finder to estimate when you are pointing directly at the Sun.

Hold a piece of white card about 15 cm behind the eyepiece and you should see a brilliant patch of light on it. If not, one trick for finding the Sun is to hold your card close to the eyepiece – as you near the Sun, the image will brighten.

Your first solar image will probably be out of focus – just a circular patch. Focus as usual, winding the eyepiece away from the telescope until the disk is sharp. That's it – you have an image of the Sun, hopefully with sunspots. At this point someone usually says, "But how can you be sure that's the Sun – the light is coming from a circular hole, and aren't those specks just dust?" It's easy to check. Change the focus – you won't get a sharp image any other way. And if you rotate the eyepiece the sunspots will stay still while any dust specks (which are usually defocused) will turn with it.

To make the image larger, hold the card farther from the eyepiece and refocus. It gets dimmer as it gets larger, and there comes a point

▼ *How to set up for projection. After making sure the finder is covered, align the telescope on the Sun using its shadow only (left). Hold a white screen some distance beyond the eyepiece and you should see a white fuzzy disk (center). Focus this to get your sharp solar image (right), on which you will hopefully see sunspots.*

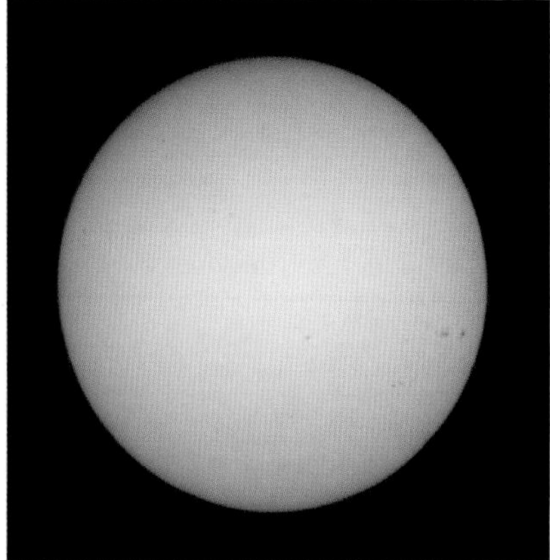

▲ *A projection screen being used on a 75 mm refractor (above left), now rare on new models. To make your own you will have to arrange a clamp around the focusing mount to hold the screen.*

Above right is the Sun as it appears on the projection screen on a typical day. A few sunspots are visible, in particular several active areas on the right-hand side.

where the light from the surroundings overpowers the image anyway. I leave it up to you to figure out how to shield the image from direct sunlight, but you can usually get a good view quite easily.

So what's to be seen? Even if there are no spots on view, you'll notice that the edge of the Sun – the limb – appears slightly darker and redder than the center. This limb darkening is a result of the Sun's atmosphere, which absorbs light, because we see the limb through a thicker slice of atmosphere than the center. If the seeing is good, you may see some mottling of the surface, called granulation. This is the result of convection cells of heat rising from below the Sun's surface layers, like those in a heated pan of water.

Usually there are sunspots. These are darker regions where magnetic fields break through the surface, restricting the light output. They change from day to day, partly because the Sun rotates but also because they are in constant motion, though you are unlikely to see any changes as you watch. They often appear in pairs, with several subsidiary spots surrounding the pair, that generally lie parallel to the equator.

Solar observers plot the numbers, positions and appearance of sunspots. Numbers vary from month to month and also over the long term, as the Sun's activity varies over a roughly 11-year cycle. When it is active, there are plenty of sunspots and there is also a good chance of seeing the aurora at night – the northern or southern lights, which are caused by streams of particles from the Sun hitting the Earth's upper atmosphere over the poles. But at solar minimum, spots can be rare. Plotting the ever-changing cycle of activity is an opportunity to do some astronomy by day, and which doesn't need a large telescope. Even a 50 mm telescope is adequate for making sunspot observations.

Drawing the solar disk is a good way to keep a record of what you have seen. It helps to have a fixed assembly on the telescope to hold the screen in place. These used to be standard accessories for refractors, but they are less common now. The idea is to project the Sun to just fill a circle of a standard 150 mm diameter. Rather than try to sketch in the spots on the screen itself, make a faint grid within your standard circle, and simply transfer the coordinates of the features you see to a report form beneath which is a similar grid, this time with bold lines so they show through the paper. It may be a low-tech solution, but it works!

As well as sunspots there are bright features called faculae. These are most noticeable near the limb of the Sun and are associated with sunspots, though there are others seen near the poles where sunspots never appear.

One thing you won't see when looking at the Sun by projection is the dramatic flame-like features at the edge of the Sun called prominences. Despite their fiery appearance, these are comparatively faint and can only be seen either when the bright part of the Sun is blotted out, as during a total eclipse of the Sun, or by using a special filter that only transmits the light of hydrogen gas (see page 154). The rippling that you can often see at the edge of the Sun is nothing more than an effect of the seeing, which can be quite poor during the day. Solar observers often find that the seeing is best in the early morning, before the ground has had a chance to heat up.

Deep-sky objects

For many people, the deep sky is the beginning and end of astronomy. The term covers such objects as star clusters, nebulae and galaxies, though the stars themselves are not generally included. More often than not, it is taken to mean the most obscure and faint objects – collectively termed "faint fuzzies," which is a good description.

To an outsider, the craze for deep-sky observing might be hard to understand. We see photos of swirling colorful nebulae, beautiful spiral galaxies and sparkling star clusters, yet when we learn that the reality is faint, lacking in detail and usually colorless, it might all seem very pointless. But when it comes down to it, many human activities are pointless. What does it matter who wins a match, who catches the largest fish or who marries whom in some soap opera?

Deep-sky catalogs

Deep-sky objects are often referred to by catalog numbers, of which the most common are M numbers and NGC numbers. The M stands for Messier, after the 18th-century French astronomer Charles Messier who listed some of the brightest deep-sky objects to help him in his search for comets. There are 110 Messier objects, and they form a convenient collection for virtually any observer.

The NGC objects are usually fainter, though there are a few bright ones that Messier missed out, the most notable being the Double Cluster in Perseus. The initials stand for the *New General Catalog* produced by the Danish astronomer J. L. E. Dreyer in 1888, to which was subsequently added the *Index Catalog*, which gives rise to IC numbers. There are 7840 NGC objects of various types, numbered roughly in order of right ascension. Many are too faint to be easily observable, though a 300 mm telescope should show virtually all of them given the right conditions.

There are many other catalogs in use, a recent example being the *Caldwell Catalog* devised by British amateur astronomer Sir Patrick Moore and published in *Sky & Telescope* magazine for December 1995. It collects 109 objects, almost all of which are also in the NGC, as a sort of extension to Messier's catalog. Then there are specialist catalogs of different types of object, such as the *Abell Catalog* of galaxy clusters. Deep-sky observers love to find yet more obscure catalogs of objects so they can impress one another with their background knowledge.

People are adding popular names to objects all the time, usually from an imagined resemblance to an object, such as the very appropriate North America Nebula (NGC 7000). The aim with some of them seems to be to get your own pet name for an object generally recognized.

So let's accept deep-sky observing on its own terms: the quest to see ever fainter and more challenging objects, even if they are not visually particularly breathtaking.

There is beauty there for sure, a subtle beauty that depends as much on the imagination as on the eye. That faint smudge is a galaxy millions of light years away. The sprinkled stars in a little-known cluster are all suns in their own right, possibly hosting alien worlds. And you have witnessed them: the photons of light that left them centuries or eons ago have traveled across vast reaches of space simply to fall into your eyes, their journey's end.

Clusters: jewels of the sky

The easiest of deep-sky objects to observe are the star clusters, which are virtually all in our own Galaxy. These fall into two types – open or loose clusters, and globular clusters. The brightest and best-known clusters are all open clusters – delights such as the Pleiades and Double Cluster. There is no regular shape to them, and they consist of dozens of hundreds of stars loosely gathered under their mutual gravitational attraction. They are easy to see because stars, being points of light, can be seen even in poor sky conditions.

Some clusters, notably the Pleiades, are so large that they don't fit in even the average low-power eyepiece, but most are smaller, such as the famous Jewel Box Cluster near the Southern Cross, which is a disappointment in binoculars, for example. Clusters can be a lovely sight, particularly in a dark sky when there seem to be hundreds of stars right at the limits of vision. Some appear to have shapes, such as NGC 457 in Cassiopeia, which has two lines of stars that have resulted in its nicknames of the ET or Owl Cluster, or the Wild Duck Cluster, which has a V-shape of stars like a flock of wild ducks. Most open clusters are to be found along the line of the Milky Way.

▲ *This star cluster, NGC 457, has various fanciful names. To understand why, you will have to observe it yourself, when the outstretched lines of stars become obvious.*

There is little to be gained by using filters to observe star clusters of any type, but it can be worth increasing the magnification somewhat, which will have the effect of darkening the sky background though the stars should remain roughly the same brightness despite the power.

The other type of cluster, the globular cluster, is more widely distributed than the open clusters. This is because they are arranged in a halo around our own Galaxy, rather than lying along its plane. However, in the northern-hemisphere winter skies, when Orion is high in the sky, there are almost no globular clusters in the sky at all. But there are some other parts of the sky that are otherwise quite barren yet which contain globulars.

Most globular clusters are comparatively remote. They appear as balls of stars, though often all but the brighter stars are hard to make

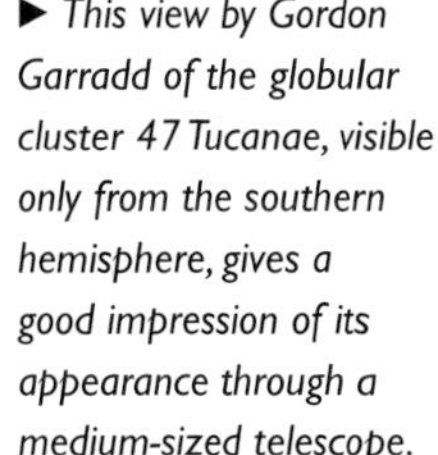

► *This view by Gordon Garradd of the globular cluster 47 Tucanae, visible only from the southern hemisphere, gives a good impression of its appearance through a medium-sized telescope.*

◄ *You need good, dark skies to see the North America Nebula, but it is easy to photograph under less than perfect conditions. The bright star at center right is Deneb, in Cygnus.*

out with small telescopes, making them look like rather grainy comets. In larger telescopes in good skies they can look spectacular. Northern-hemisphere observers have to put up with some fairly poor specimens, such as M13 in Hercules, sometimes called the Great Cluster. But it can't compare with the finest globulars, which are in the southern hemisphere – Omega Centauri (visible low in the sky in much of the US) and 47 Tucanae.

However, I suspect that interest in globular clusters rapidly fades as they get fainter, probably because one looks very much like another, even though they may be easier to see than many of the other objects that are on the lists. So let's hear it for the globular clusters. The larger the aperture the better you will see them, but they are still worth looking for even with small apertures.

Nebulae: welcome clouds

Most observers detest clouds, but the nebulae are a different matter, though the name is Latin for "cloud." Prior to the early 20th century, when the true nature of astronomical objects became known, almost every fuzzy object in the sky was called a nebula, but now the term is reserved for true gas clouds. There are five basic types of nebula: hydrogen or H II regions, reflection nebulae, supernova remnants, planetary nebulae and dark nebulae.

◄ *The Eagle Nebula, M16, in Serpens. Although the visual view shows no color, signs of the dark pillars in the center can be seen using amateur telescopes.*

▲ *The blue reflection nebulosity surrounding the Pleiades shows up well in photos, but it needs very dark skies to be visible convincingly to the eye.*

The brightest nebula in the sky, the Orion Nebula (M42), is an example of a hydrogen or H II (pronounced H-two) region. The H II refers to the fact that the hydrogen is caused to glow by the light from nearby stars (in technical terms, it is ionized), which sets it apart from the dark H I gas. The predominant color of an H II region in photographs is red, but the eye is insensitive to this color and most nebulae appear either colorless or green, no matter what color they appear in photos.

Because H II regions emit light that consists of specific colors or wavelengths only, they are just the sort of objects that light-pollution rejection (LPR) filters are designed for. Unless you have a virtually perfect sky, a filter will help to darken the sky background and help you to observe the nebula. But from a heavily light-polluted area, even filters will not make the fainter nebulae visible.

Nebulae vary considerably in size, some being so large that they can be seen only in binoculars, and then with difficulty. It helps to use an atlas or catalog that shows the size of the nebula, so you know what eyepiece to use. But if H II regions are hard to see, particularly from city areas, they are often easy to photograph, especially using red-sensitive film.

Reflection nebulae are dust clouds that are illuminated by a nearby bright star, and they appear blue in photographs – not because dust is blue, but because this is the color that is most readily scattered by dust. As a result, LPR filters should not be as useful as on H II regions, though they should help.

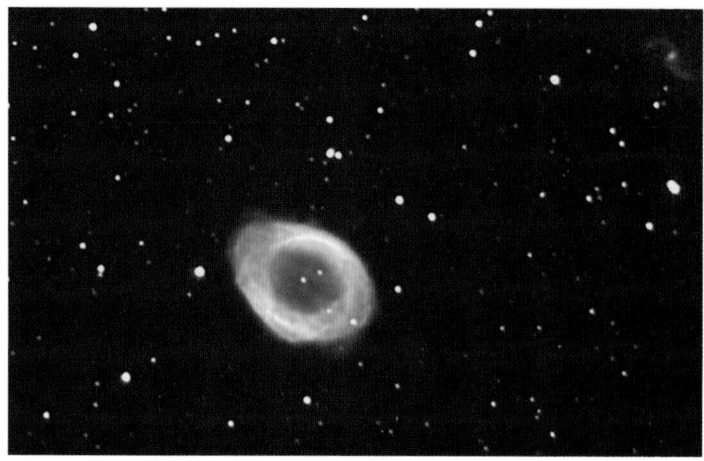

▲ *"Doughnut Star" is how a young girl described the Ring Nebula to me on being shown it through a telescope. It is one of the most popular planetaries in the sky, though the 15th-magnitude central star is a great challenge.*

There are only a few supernova remnants available to the amateur, the best known being the Crab Nebula (M1) and the Veil Nebula (NGC 6960 and 6992). The Crab is fairly compact and none too easy to see in suburban skies, while the Veil is a circular tracework of gas that is visible even in average skies with a medium-sized telescope and a suitable filter.

Planetary nebulae are among the most extraordinary objects in the sky, and they are the only deep-sky objects to show any color – some of the brighter ones can be blue or green. They are the shells surrounding dying stars, and frequently the star itself is visible at their centers. They are not necessarily spherical, and some have quite unusual shapes. They are often very small, and can appear starlike unless you use a high power. But being condensed, they can be easier to see than many large H II regions. The Dumbbell Nebula in Vulpecula is large enough to be visible in binoculars, while the Ring Nebula in Lyra is easily seen in a small telescope using a magnification of 25 or more.

At 76 arc seconds diameter, the Ring Nebula is actually larger than any planet, but it is the disk-like nature of many planetaries which gave them their name. Despite being the same general size as planets, their comparative faintness means that you need far higher magnifications than you would need to see detail on a planet. As with H II nebulae, LPR filters can be a great help.

Dark nebulae can only be seen in silhouette against a background of stars or a nebula. The best-known example is the Horsehead Nebula in Orion, but the Great Rift in Cygnus and the Coal Sack next to the Southern Cross are also dark nebulae.

▲ *The Horsehead Nebula is a dark nebula within the pale glowing nebulosity south of Zeta Orionis, the most southerly of the three stars of Orion's belt. The nebula to the left of Zeta is called the Flame Nebula.*

Gorgeous galaxies

It is probably the nature of galaxies that makes them so popular – far beyond our own Galaxy, they stretch away into the Universe so far that their light has taken millions of years to reach us. They have a variety of shapes, but in small telescopes all one often sees is a roughly circular blob, which is the galaxy's nucleus. A case in point is the famous Andromeda Galaxy, M31. You can see this with binoculars even from city centers (given a suitable park and shielding from bright lights), but all you can see is the central region. In dark skies, the nucleus appears much larger and is surrounded by a larger region which is the spiral arms. At least one dark band is visible, being one of the dust lanes in the arms.

There are three main types of galaxy visible to the amateur: spirals, ellipticals and irregulars. The spirals are undoubtedly the

► *With binoculars or a small telescope the Andromeda Galaxy (upper right) looks like an oval blur. The central nucleus is much brighter than the surrounding spiral arms, just visible in this view.*

most popular, particularly those that are face-on, so their spiral arms are well separated. But you do need a largish aperture to see them properly. Quite often you imagine you can see the arms of a galaxy such as M51 because you know what you are looking for, though it took Lord Rosse's famous 72-inch mirror in Ireland to reveal them for the first time. Admittedly, modern telescopes are much more efficient, but it is hard to see a galaxy such as M51 with completely fresh eyes.

Some spirals are seen edge-on, so they appear spindle-like, and these too are firm favorites because the thin cigar-shape is unlike most other objects in the sky. Sometimes they have dust lanes along their middles.

Elliptical galaxies are just that, with no spiral arms. They are usually quite featureless, but they vary from circular to quite elongated ellipses. There may be a nucleus that appears quite starlike – don't mistake this for a supernova, but check before alerting everyone to your momentous discovery.

Irregular galaxies are usually fairly faint, except for the glorious Magellanic Clouds in the southern hemisphere, which are companions of our own Galaxy and are easily visible with the naked eye. Some galaxies are irregular because they are interacting with each other over enormous timescales: tidal forces drag out streamers of stars that are visible as faint wisps of light.

As the light from galaxies is basically starlight, LPR filters are of limited use, since they dim the galaxy as well as the light pollution.

Seeing double

Double stars are usually true pairs of stars that are orbiting each other. A few of them are optical doubles – that is, they just happen to be in the same line of sight – but in fact a large proportion of all stars are double. Many are too close together to be seen as separate stars, but each constellation has its crop of easily observed doubles.

There are a few classic showpieces in the sky, almost everyone's favorite being Beta Cygni or Albireo, which can be seen as two stars of yellow and blue even in the smallest telescope. The advantage of double stars is that many are bright enough to be seen in even the worst skies and smallest telescopes, but after the first few it has to be said that

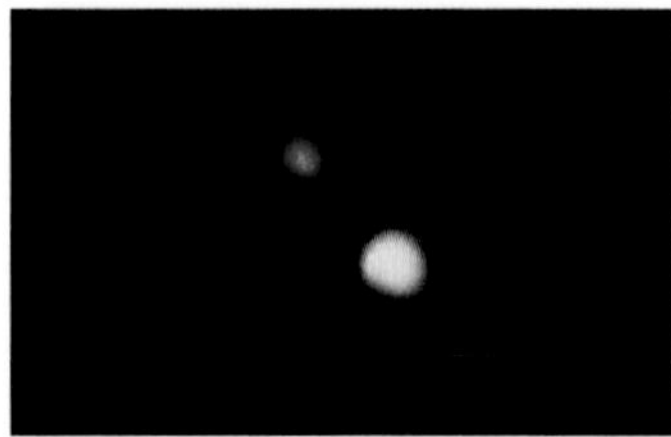

◀ *Arguably the most beautiful double star in the sky, Albireo (Beta Cygni) is bright enough that it shows its colors just as well to the eye as in a photograph.*

many people have seen enough of them. The prettiest are those with contrasting colors, and some doubles have been described with quite fanciful colors such as orpiment and sardonyx (actually subtle shades of yellow and red). But there is a further use for double stars – as test objects, if you know that a particular pair have a similar separation to the resolving power of your telescope. To be a fair test they should be of roughly similar brightness and a few magnitudes brighter than the limiting magnitude of your instrument.

A few people make painstaking and accurate measurements of the separation and angle between double stars, an occupation that has some real scientific use, but this is not for most observers, who are content to admire a few favorite doubles on each session, and maybe find some they are unfamiliar with.

Stars that vary

Not all stars are constant in brightness. Some are variable stars, whose output changes over a period from minutes to centuries. Monitoring the light output of variables is quite easy to do, requiring only a telescope and that excellent light detector, the Mk 1 eyeball. The trick is to compare the brightness of the variable with other stars whose brightness is known to be constant. There are several ways of doing this, and the website of the American Association of Variable Star Observers gives guidelines on how to go about it.

For each variable you observe, you need a chart showing the stars to compare it with, and many of these are also available online from a variety of sites. The Society for Popular Astronomy has online charts for some bright stars that are not on other organizations' programs, that can be observed with binoculars or even the naked eye. This is one field of observation where a high-power telescope is not usually needed – the main thing is to be able to see the stars over a fairly wide area of sky at a glance, so you can compare the brightnesses. And the results are of real use to professional astronomers, who do not have the resources to monitor hundreds of stars every night. Variable-star observing may not be glamorous, but it is certainly worthwhile.

www.aavso.org

www.popastro.com/sections/vs.htm

7 • BUYING MORE: A GUIDE TO ACCESSORIES

Astronomy magazines are full of ads – not just for telescopes, but for extras. You can spend as much on these as on the telescope itself. I suspect that most people are cautious about acquiring too many bits and pieces, and want to be sure they need them before buying. But there are some extras that are definitely worth considering. Top of the list are extra eyepieces, which are really the only essential extras, then adapters for astrophotography, light-pollution filters and maybe devices for reducing the problem of dewing. The telescope market is a growing one, with new products becoming available all the time, so I can't give a complete guide here. There is no substitute for asking around for help, either among members of your local astronomical society, through online newsgroups and forums, or even from manufacturers.

Eyepieces

Many telescopes are sold with only one eyepiece, often a 25 mm or 26 mm, which is adequate for finding objects in the first place and for enjoying views of the larger objects such as star clusters and nebulae. But a telescope with a single eyepiece is rather like a car with only one gear. A range of two or three is really needed – one for general purposes and finding objects (power 60–100), one for medium powers (100–200) and maybe one for high powers (200–350, if your telescope will stand it). The last one may well get the least use if your observing site is plagued by bad seeing. If you want to be cautious, see how you get on with a medium power before going for the high power.

The exact focal lengths you need depend on the focal length of your telescope and the range you want. Plan your range carefully, so

◄ A selection of eyepieces of about 9 mm focal length and therefore giving similar magnification on a particular telescope. At the rear is a 9 mm Nagler giving an apparent field of view of 82°; in front, from left to right, are a Meade Super Plössl (52°), a Meade Plössl (50°) and an unbranded 9 mm Kellner (40°). All are in the $1\frac{1}{4}$-inch fitting, though the Nagler also fits 2-inch focusers.

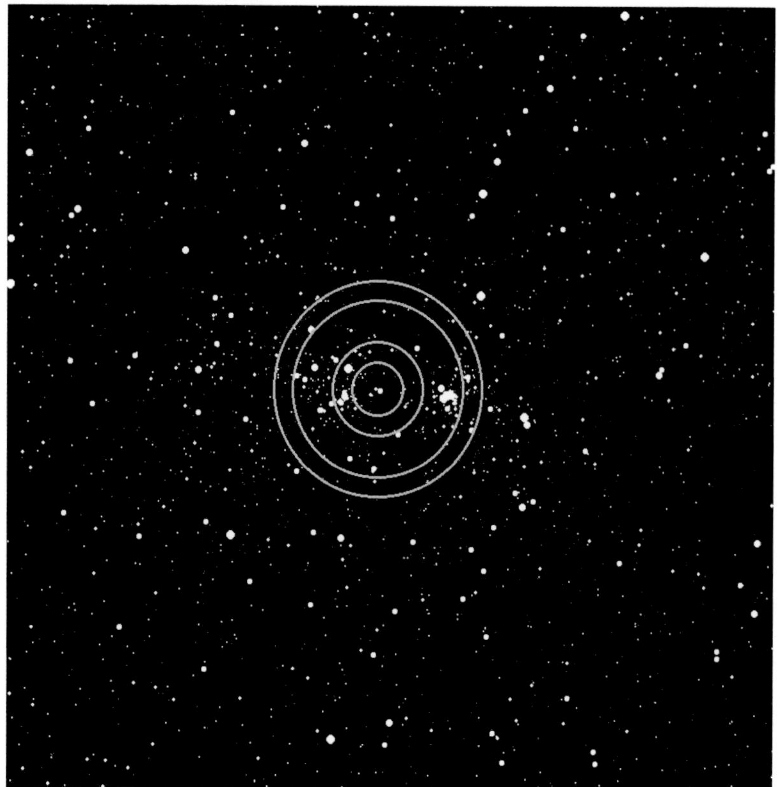

▲ *The Double Cluster in Perseus with the actual field of view of various eyepieces as used with a typical 200 mm f/10 SCT. From the outside: 20 mm Nagler (actual field of view 49 minutes); Meade 26 mm Super Plössl (40 minutes); standard Kellner 18 mm (20 minutes); and 9 mm orthoscopic (12 minutes). The diagram was made using the SkyMap program, which displays eyepiece fields.*

as to get maximum use out of each eyepiece. You might think that zoom or vari-power eyepieces would be the norm, as they are in the case of photographic lenses, but the traditional wisdom is that the extra internal glassware these involve can introduce unnecessary distortions. However, some very highly regarded eyepieces have eight or more elements (individual components), so sheer volume of glass should not be a problem. In fact, the latest zoom eyepieces from recognized manufacturers are reputed to be very good, offering a range of focal lengths between about 8 mm and 20 mm. The cost is equivalent to about three good-quality eyepieces, and of course the power is variable

Types of eyepiece

Huygenian – The simplest and cheapest eyepiece type, often sold as the standard eyepiece with department-store refractors. Has a small field of view and only gives acceptable results with telescopes of focal ratio longer than about f/10. But because these eyepieces are so cheap and simple, they are well suited for projecting the Sun, as there is no cement between the elements that may otherwise melt. If you wish to project the Sun using a reflector, stop the telescope down so that the aperture is reduced and the focal ratio is increased, and you may get a satisfactory image using an old 25 mm Huygenian without risking your expensive eyepieces.

Kellner – Another low-cost eyepiece, with a comparatively wide field of view. Often used as binocular eyepieces. Suffers from ghosts and other aberrations, but gives acceptable results with faster instruments. Dust on the field lens can be a problem as it is close to the focus point. Meade Modified Achromat eyepieces are based on the Kellner.

Orthoscopic – For many years regarded as the ideal eyepiece, with a field of view of about 45 degrees fairly free from troublesome aberrations. A few manufacturers still sell them.

Plössl – Now the standard eyepiece, with excellent characteristics, and capable of giving wide fields even with fast telescopes. There are several different designs all with the same general name. But cheap Plössls imported from the Far East may give disappointing results.

Extra wide field – A number of modern designs, with fields of view up to 84 degrees. Noted for their excellent performance, but heavy and expensive. Examples include Meade Ultra Wide Angle and Tele Vue's Nagler eyepiece.

SOME TYPICAL EYEPIECE SPECIFICATIONS

Type	Focal length (mm)	Field of view (degrees)	Weight (grams)	Eye relief (mm)
Huygenian	9	42	35	3
Kellner	10	43	43	9
Orthoscopic	9	40	31	7
Plössl	8	50	45	6
Lanthanum	9	50	158	20
Erfle	20	65	100	10
Nagler	9	82	410	12

so you can choose your magnification according to the conditions, although not all zoom eyepieces stay in focus as you change power.

Another drawback is that the apparent field of view (the size of the circle of light that you see) is smaller than that of a conventional eyepiece of the same focal length. This, however, is no drawback if you mainly observe planets or the Moon. Do not confuse the apparent field of view with the actual field of view – the amount of real sky you see. This depends on the focal length of the telescope, and is usually much less than a degree.

What the figures mean

Eyepieces are always described by their focal length, which is inscribed on them. It is not necessarily obvious where the focal point lies and is measured from – you can't measure it by forming an image on a wall, as you can with an objective.

A specification that is not always quoted is the eye relief – the distance the eye must be from the eyepiece in order to see the whole field of view. The shorter the focal length, the less the eye relief, but some designs have much longer eye relief than others for a given focal length. Eye relief matters particularly if you wear glasses for observing.

Another figure to bear in mind is the exit pupil. This is the diameter of the beam coming from the eyepiece, and this too becomes smaller as focal length goes down. However, all eyepieces of a given focal length have the same exit pupil with a given telescope. So what is the point of knowing it? It can be important to work out its value when considering buying a low-power eyepiece. Find it by dividing the eyepiece's focal length in millimeters by the focal ratio (the f-number). So a 26 mm eyepiece on an f/10 telescope has an exit pupil of 2.6 mm. On an f/5 instrument the same eyepiece will have an exit pupil of 5.2 mm. But use a 40 mm eyepiece on the f/5 instrument, in order to get a wider field of view, and the exit pupil becomes 8 mm.

Now the problem is that the eye doesn't usually open as wide as that, so not all the light from the eyepiece will enter your eye. The maximum, when dark-adapted, is often quoted as being 7 mm, though in people over 50 it may be 5 mm or even less. So you might as well have bought a telescope of smaller aperture in the first place. However, you might be prepared to sacrifice a bit of light in order to get a dramatically wide field of view. It is often said that 7 × 50 binoculars, with an exit pupil of 7 mm, are wasted on people whose pupils do not open as wide as this.

Some eyepieces are described as parfocal. This means that all the eyepieces in a range are designed so that they all have the same focus position, or very nearly, so refocusing is kept to a minimum when changing eyepieces. But there is no standard, so one range of parfocal eyepieces may not have the same focus point as another range.

If I had been writing this a few years ago, I would have described the traditional range of eyepieces, with their strengths and weaknesses – the Ramsdens, Kellners, orthoscopics and so on. But in recent years, these types have all but disappeared from the mainstream market as manufacturers have gone over to what was once a comparatively rare type – the Plössl. These, plus various types of super-wide-field eyepiece such as the Nagler, now make up virtually the entire range.

There are several things you have to look for in an eyepiece. First, you have to decide on the focal length, then you must decide whether or not you need, and are willing to pay for, a wide field of view.

Many beginners buying their first additional eyepiece not surprisingly go for one that gives a fairly high power. The favorite is around 9 mm focal length, giving a power of 200 on a 200 mm SCT or around 150 on a smaller instrument. After being used to the view with the standard eyepiece you may be surprised by how dim the image appears when using a 9 mm and how hard it is to find the object again, having changed the eyepiece. There is no denying the appeal of a high power, but observers with a choice probably use their 9 mm less often than they do an eyepiece of 13 to 16 mm.

An option is to buy a Barlow lens. This multiplies the power of any eyepiece it is used with by a standard factor, usually either 2× or 3×, though a 5× Barlow is now available. So a 2× Barlow turns a 26 mm eyepiece into a 13 mm, and a 16 mm into an 8 mm. Some people prefer to have a single eyepiece for each focal length, but a Barlow and two other eyepieces give you a choice of four focal lengths, if you choose sensibly. The 5× Barlow is intended for use with short-focus instruments, with which it can be hard to get high magnifications using conventional eyepieces – but you need to be sure that your telescope is of good enough optical quality to withstand such high powers in the first place. It is also useful for increasing the effective focal length when imaging planets.

One advantage of a Barlow is that you retain the eye relief of your original eyepiece (see box on page 147), even when it is working at a higher power. This is useful for anyone who wears glasses when observing. But it is important to get a good Barlow from a recognized manufacturer, otherwise it will simply degrade the performance of any eyepiece it is used with.

Another type of eyepiece that is renowned for its good eye relief is the Lanthanum type. Lanthanum is a rare earth element used in glass to change its refractive index, permitting improved designs. Someone who needs to use glasses when observing because of astigmatism may find Lanthanum eyepieces very helpful, but the improved eye relief tends to come with poorer optical definition.

Field of view is the holy grail of eyepiece manufacturers. At one time, eyepieces had what amounts to tunnel vision – very small apparent fields of view. Over the years, fields of view have improved until now one can get amazingly wide fields, almost like picture windows. And unlike older wide-field eyepiece

◀ A 2× Barlow lens between two eyepieces – a 25 mm and a 9.5 mm Plössl. This combination will give you four magnifications suitable for an f/6 telescope – the highest power would probably be excessive with a longer focal ratio.

types, such as the Erfle, the definition is crisp almost from edge to edge, and they are available in a wide range of focal lengths, from 40 mm right down to less than 5 mm, though not every model has identical performance. These eyepieces can give amazing views, and their owners often use them as the eyepiece of choice for every purpose, including planetary work where wide field of view scarcely matters because the object is very small within the field of view, even using a high-power eyepiece. Anyone with a telescope of short focal ratio, such as f/4, will find that the more conventional types of eyepiece will not give good definition at the edges of the field of view, so to get the best from their telescope they will have to splash out on an expensive eyepiece.

But before deciding that you must have one of these, find out whether it will fit your telescope. Some of the longer focal lengths are available only in a 2-inch eyepiece fitting. Most eyepieces sold these days have 1¼-inch push-fit barrels, though those for some small department-store instruments still have the Japanese 24.5 mm (0.965-inch) barrels. The optical performance of the smaller eyepieces may be identical to their larger-barreled counterparts, by the way, so do not discard them on the basis of their size. Some larger telescopes have the 2-inch fitting, requiring an adapter for the standard 1¼-inch eyepieces, and most SCTs can take either 1¼- or 2-inch eyepieces.

These ultra-wide-field eyepieces not only cost as much as some small telescopes, they are also very heavy, weighing about as much as a camera body. On a Newtonian you will almost certainly have to rebalance the tube in order to use one. It is also quite likely that a Newtonian reflector will require a larger size of secondary mirror to cover the full field of view of a wide-field eyepiece. But with short-focus reflectors there is an additional problem with very wide-field eyepieces: the secondary is so large that it becomes an obstruction in the field of view, and you see an annoying black blob at the center. Also bear in mind the matter of the size of the exit pupil (see the box on eyepiece specifications on page 146): if it is larger than the pupil of your eye you will not be using the full aperture of your telescope, though you may be prepared to sacrifice a little brightness in return for the wide field.

Are wide-field eyepieces worth the money? Their fans point out that they are great for observing some nebulae, nearby galaxies and comets, but they also have other advantages. Their wider field of view makes it easier to find objects, particularly if you don't have the advantages of computer control and setting circles. Some galaxies have nearby companions, and the view of several objects together is rewarding. Even the sight of a lonely galaxy against a backdrop of stars is a spectacle in itself. You can view the Moon as if from a spaceship, particularly with a high-power eyepiece. For planets they are not necessary, nor are they for planetary nebulae and double stars. But observational astronomy is largely about spectacle, and wide-field eyepieces certainly provide that.

The width of an eyepiece's field of view is not its only attribute. Other factors you have to consider are absence of curvature of field, in which the eyepiece has to be refocused for different parts of the field of view; freedom from ghosting, which means spurious ghost images of bright objects, particularly planets, floating around in the field of view; and of course the lack of various aberrations, meaning distortions of the image, particularly near the edge of the field of view. Some eyepieces have additional coatings on the optical surfaces to improve the light transmission. It may seem odd that actually coating a layer of material on glass results in more light going through, but in fact the aim is to reduce the amount of light reflected off the surface of the glass, which can be considerable in uncoated glass. Good eye relief is also important (see box on "What the figures mean" on page 147).

Filters

Next to eyepieces, filters are high on an astronomer's wish list. If only the perfect filter existed! Actually, some are useful and some are not. They divide into four types – color filters, solar filters, light-pollution filters and minus-violet refractor filters. All except the solar filters screw into the telescope side of the eyepiece, which should have a standard thread, so you need to make sure when buying any filter that it is the

▲ *A threaded eyepiece filter (in this case, a light-pollution filter) and two plastic color filters intended for photography.*

▲ *Beware of Sun filters supplied with small refractors! This one actually did crack when in use, though fortunately no-one's eyesight was damaged.*

right fitting for your eyepieces. Some similar filters are made to fit within the focusing mount of an SCT, for use in photography, about which more shortly.

You may come across a solar filter that screws into the eyepiece of small imported telescopes, but these should never be used. They are notorious for cracking as they absorb the Sun's heat. The only safe place for a solar filter is over the top end of the telescope, covering the full aperture, where it is only subject to normal sunlight.

Color filters

Most observers survive perfectly well without ever using a color filter. They are of no use at all when observing deep-sky objects, and are only worth considering for lunar and planetary observation. Their value lies in enhancing the faint markings on planets by using a color filter complementary to the color of the marking. A bluish marking will be emphasized by a yellow or orange filter, and vice versa. A dark or neutral density filter will reduce the brightness of the full Moon, making it less painful to look at when you are dark adapted, but is not essential.

There is a wide range of filters available, but it would not be a good idea to go out and buy every one on the basis that it might come in handy – there are cheaper, if less convenient, alternatives.

SOME FILTERS AND THEIR USES			
Wratten no.	**Color**	**Transmission**	**Uses**
8	pale yellow	83%	Markings on most planets
12	deep yellow	74%	As above. Improves daytime visibility of Mercury and Venus
38A	deep blue	17%	Markings on planets, particularly Great Red Spot
80A	pale blue	30%	As 38A, but for smaller instruments
56	pale green	53%	Markings on Mars in particular

Transmission is the percentage of all light that the filter allows through.

Visit a camera shop and look at their range of square plastic filters, designed to sit in a special adapter on the front of a camera lens. For a fraction of the cost of one official eyepiece filter you can buy several plastic squares in a range of colors. You can experiment with these simply by holding them in front of the eyepiece. Then if you decide that you need a proper eyepiece filter you can buy one.

The numbers by which many color filters are described date back to the turn of the 20th century when an English chemist named Wratten introduced a range of filters. His company was bought out by Kodak, and today the numbers are used (often preceded by a "W") by most manufacturers, though only Kodak make the "official" Wratten filters, available as squares of gelatin. These are quite convenient, as you can cut the gelatin to fit inside eyepieces and lenses, but these filters are not particularly cheap.

Solar filters

The important message when observing the Sun is safety first. The Sun is blindingly bright, and the least dangerous way to observe it is by projecting its image, as described on page 133. But this is not always convenient, and some solar features can't be seen on the projected image. So if you use a filter, make sure that it is firmly fixed over the top end of your telescope so no direct light can get through, and that there is no risk of it coming loose.

There are two basic types of solar filter – those that simply cut down the Sun's light and heat, and those that only transmit specific wavelengths of light, notably the hydrogen-alpha wavelength.

General solar filters are made from either glass or plastic film, such as Mylar, coated with a thin layer of metal particles to block light.

Aluminum is the most common coating, giving a bluish transmitted image, but there are others, such as Inconel, with different colors. Another variety uses black polycarbonate, in which the absorbing particles are spread throughout the material. This gives an orange color to the Sun, which is regarded as more natural, even though the Sun is really white so no coloring is correct.

These filters are carefully manufactured with the correct density of coating for their purpose. The important thing is that they should absorb light equally over the wavelength range. It is not just the visible light that causes eye damage – the invisible ultraviolet and infrared rays can be equally harmful, but unlike the visible light, you can't tell if these are being absorbed. I have seen bits of plastic sold as solar filters that were virtually transparent in the infrared. The Sun looked dim through them, so the eye happily allowed you to keep on looking, whereas in the case of a bright image the eye-brain combination instantly forces you to look away. As a result, the infrared rays could cook your retina unnoticed until it was too late.

Mylar-based filters must have a double coating, so that any pinholes in one layer caused by creasing will hopefully not line up with those in the other layer. They should not be stretched tight but should be allowed to sag and be slightly wrinkly. This does not look very neat but it avoids strain on the filter. The material is so thin that the wrinkles have no measurable effect on image quality.

Glass filters are more expensive than the Mylar variety. Whatever the material, it has to be firmly mounted in a ring that fits neatly around the top end of your telescope. These are made for standard telescope designs, but if you have an old or unusual telescope you will have to get one made to fit.

The other type of filter is more selective in the wavelengths it allows through. The pink hydrogen prominences on the

► *A Meade ETX-105 fitted with a full-aperture solar filter from BC&F Engineering. This coated glass filter gives the Sun a pinkish hue.*

Sun's limb emit light at specific wavelengths, notably the one known as hydrogen-alpha or H-alpha, which is 656.3 nanometers (abbreviated "nm"). By using a filter that transmits only around this wavelength, the light from the rest of the Sun is cut out leaving only the prominences and various other features on the Sun's surface. The image is deep red in color, the color of this wavelength. The narrower the passband, the better you will see the features on the Sun's surface, though a fairly wide passband is adequate for viewing prominences. Typical filters have a passband of only 0.1 nm, often quoted in the more traditional Ångstrom units of wavelength, where 1 Å = 0.1 nm.

Comparatively wideband filters will show the prominences and cost about the same as a Meade ETX-90 Maksutov telescope. They consist of a full-aperture filter that cuts down most of the energy entering the telescope, and the main filter that sits in front of the eyepiece. The more expensive narrowband filters have to be heated to the correct temperature, so they require an external power source – and also a large budget, as they can cost as much as a computer-controlled SCT.

Incidentally, if you see what seems to be a cheap H-alpha filter for sale, it is probably of the type used for deep-sky photography rather than for solar observation. These have a comparatively wide passband and are not suitable for solar observing, though they are excellent for their intended purpose.

Purpose-built solar telescopes are available with built-in filters. An economy model of such telescopes is also available, the Coronado PST, which for the optical tube alone costs about the same as an ETX-90 Go To telescope. But enthusiasts delight in viewing the ever-changing solar disk, and these instruments have recently become very popular.

Light-pollution filters

Every astronomer must have wished for a filter that simply cuts out the yellow glare that pervades most of our skies these days. Streetlights, home and office lights, advertising signs, car lights and hundreds of other sources create a glow in the sky for many kilometers around an urban area. What looks like pleasant, remote countryside by day turns into an extended suburb, as far as lighting is concerned, by night. Few places are completely immune from this. Most professional observatories can detect the lights from cities even hundreds of kilometers away.

Many amateurs are simply not able to get away from the light pollution on a regular basis, so they have to find ways of trying to beat it. If you could analyse the spectrum of the light-polluted night sky – that is, the detailed make-up of the colors in it – you would see a general rainbow of color caused by what are called continuum sources, with several bright individual colors superimposed. The continuum is caused by such sources as ordinary light bulbs, security lights, car headlights and many floodlights, which give out light right across the optical spectrum. The other sources, those that give off specific colors of light, are mostly streetlights, together with some advertising signs and colored floodlights.

Sadly, almost nothing can be done about continuum sources, because they cover exactly the same part of the spectrum as the light from ordinary stars and galaxies. The only solution is to campaign for more careful lighting, so that the lights do not shine up into the sky.

The other types of light give off very specific colors. The spectrum of these is usually called a line spectrum, because it consists of bright lines rather than a whole rainbow. This makes it a little easier to filter them out, as in theory all you need to do is to find a filter that blocks just those specific colors. But this, as you might expect, is easier said than done.

One problem is that simple color filters are not precise enough in their effects to be of any use, so light-pollution filters are of a type generally known as interference filters, which have numerous thin coatings that control the transmission of light more precisely. These make it possible to make filters that block only certain colors and let through the rest. Another problem is that some of the most common streetlights, the pinky-yellow high-pressure sodium lights that are so common these days, have broader lines than most others so they emit light over much of the spectrum anyway. And finally, some of the lines from streetlights coincide almost precisely with those from objects in the sky, such as nebulae, that also have line spectra.

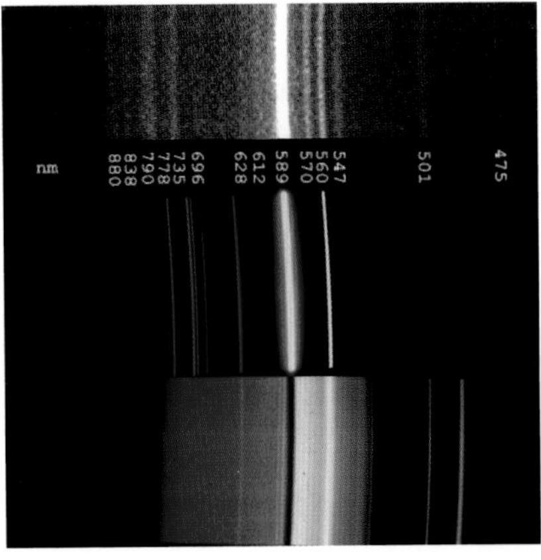

► *At top is the spectrum, in black and white, of the clear night sky, taken by Bob Neville from rural Northamptonshire, UK, showing the bright lines mostly caused by streetlighting. The intense bright line at 589 nanometers is caused by low-pressure sodium (center spectrum), but there is a broad glow on either side of this caused by high-pressure sodium (lower spectrum). The wavelengths of the night-sky and streetlight spectra do not match precisely as they were made using different systems.*

As a result, there are several different types of light-pollution filter (often called LPR filters, for light-pollution rejection). The most common, and the least expensive, cuts out a broad band of the spectrum where light pollution is strongest, and transmits light in a part of the spectrum where most nebulae emit their light. These help you to observe some objects, such as planetary nebulae, but have minimal effect on stars and galaxies. What they won't do is to turn a yellow sky into a black one. If you have a badly light-polluted sky, they will have a very small effect. But if your skies are not too bad, they may make the difference between seeing a deep-sky object and not seeing it at all. Examples include the Lumicon Deep Sky filter and the Celestron LPR filter.

Another type, known generally as narrowband filters, are designed to cut out most of the light from the sky except for the region where emission nebulae (H II regions) emit most of their light. These really do help, both in light-polluted and fairly dark skies, but again they will not work wonders in severely bright sky areas. The Lumicon UHC and Meade Narrowband Nebular filters are of this type.

The most expensive of all allow through only certain wavelengths of light, specifically chosen to transmit the light from specific types of object, such as the O III lines due to oxygen, the H-beta line of hydrogen and the Swan band in comets. These do really turn the sky black, but at the cost of transmitting only a very select band of radiation from the sky. Each one works on only certain objects. The Lumicon H-beta filter, for example, is sometimes wryly called the "Horsehead filter" because the Horsehead Nebula in Orion is one of the few objects that benefits from using this filter. Line filters tend to be good purchases to share among a group of friends or an astronomical society, rather than being "must haves" for every observer. O III filters can have an almost magical effect on some objects, but owners of small telescopes will find them unrewarding as the view becomes very dim indeed.

Minus-violet filters

Among the advantages of interference filters is that they have been able to tackle the long-standing problem that bedevils ordinary achromatic refractors – that of blue or violet fringing around images. These telescopes are now very popular, but unless you use them only at low powers or splash out on an apochromatic refractor, such fringes are inevitable. Ordinary color filters are not precise enough to improve the situation, but interference filters can cut down the fringing without seriously affecting the overall image color. A typical filter, such as the Baader Fringe-Killer, has a straw color, for example.

Finders

A good finder is worth having. If you are unhappy with the performance of the one fitted to your telescope, consider replacing it with a more powerful one. You might consider a right-angle finder, which has a 90-degree star diagonal. This gives the sky the same orientation as seen in the star diagonal of a refractor or SCT, or on a Newtonian you can mount it just above the eyepiece, although its field of view will be reversed as seen through the main telescope. Right-angle finders take some getting used to, however, and if you are still familiarizing yourself with the sky they could simply add to your confusion rather than helping.

Virtually all finders give an inverted view of the sky, like the main telescope, for no good reason that I know of other than that they have always been made like that. But to me it makes much more sense to have a non-inverting finder that matches the naked-eye view of the sky – like the binocular view, but with a crosshair. It hardly matters that the view through the telescope is inverted – what matters is to be able to find an object easily.

Currently most non-inverting finders are either rather small or have right-angle prisms, which means you don't look directly toward the object when finding it. But a few good non-inverting finders are now available, including some with variable illumination for the crosswires – the ideal situation in my view.

Alternatively, consider a zero-power finder, which simply allows you to point the telescope with the naked eye much more precisely. The original zero-power finder was the Telrad, and now you can buy other designs, such as the Orion EZ Finder. These devices won't show you the stars any better, but instead act as a super-peepsight. As you look into the finder from behind, you see a set of illuminated rings or a dot projected on to the sky at infinity. With these you can often point the telescope at a specific object just as well as with a conventional finder, as long as you can see enough nearby stars to get your bearings. But in a bright city sky, a zero-power finder will not be as useful as it is in a dark sky.

Dealing with dew

Wherever you live, dew on the optics can be a problem, but in a damp climate, such as that of Britain, it is an almost constant companion. Newtonians are comparatively free from it, but refractors and SCTs, with their exposed areas of glass at the top end of the tube, suffer badly. Water from the atmosphere will condense on any surface that has cooled below a certain temperature, and exposure to the night sky is the best way for something to cool down quickly. One way to help prevent dew is to use a long dew cap on the front end of the telescope to reduce the exposure of the glass to the night sky. Refractors usually have one fitted as a matter of course, but SCTs do not – presumably because to fit one as standard on such a wide tube would cost more. The cap should extend about twice the tube's diameter.

You can buy standard dew caps from the SCT's manufacturer, or you can simply make your own from rolled cardboard. Either way, they will not eliminate the problem of dew but will simply slow down the rate of cooling of the glass, hopefully for long enough. But if dew remains a problem, you can buy battery-powered heating tapes to fix around the inside of the tube with Velcro.

A cheaper alternative, though less convenient, is to buy a small hair dryer intended for use in a car. On cold nights the stream of air is only slightly warmer than the surroundings, but it is enough to get rid of the dew from a lens or corrector. In fact, it would be a bad idea to use too hot an air stream as this could distort the glass and in extreme cases even cause it to crack.

Portable power supplies

Many, though not all, telescope mounts these days run from a 12v DC supply, exceptions being the Synta EQ2 mount (6v DC) and the Russian TAL-1M (domestic AC voltage). Traditional sources of power, other than battery packs of AA cells which rapidly lose power, are the car cigarette lighter socket and the AC adapter, neither of which is always convenient. But 12v power packs are now available that are in theory intended for emergency use in starting a car, though how often they are used for that purpose is debatable. They are ideal for running a telescope, however, and one charge should last for many nights of observing, and will also provide onsite power for the portable hair dryer.

If such a unit is beyond your means, here's a tip: go along to your local car battery retailer and beg a used car or motorcycle

◀ A 12v portable power supply being used to run an ETX-105. A splitter cable also allows a portable hair dryer to be used.

battery. (Motorcycle batteries are easier to carry around but hold less power.) They may be happy to give you one, to save them the cost of disposing of it. A pair of crocodile clips linked to a cigarette lighter socket will give you a handy power supply. Even if the battery no longer starts a car, it will probably run your telescope and a little hair dryer for months even without a further charge.

A shot in the dark

The sights you see through your telescope are just crying out to be photographed, and in fact many of them, such as deep-sky objects, can look so much better in a photograph than you can ever see with the eye. This is because film and its digital equivalents have the ability to carry on storing up light for minutes or even hours, revealing yet fainter objects, which also makes it possible to record the colors of deep-sky objects that are beyond the eye's sensitivity. Pictures in magazines tempt you to try your skill by getting the accessories to attach a camera to the telescope.

Be warned: astrophotography through the telescope is not easy. The devices you can buy help enormously, but there is still a lot of skill and patience needed. For every one effort that works out well, there are probably ten that are failures. But the results can be well worth it, and your non-astronomical friends could be amazed at what you can manage. If you are really good, even your astronomical friends who have tried and failed will be impressed.

I can give only an outline here. There are several good books on astrophotography, so if your appetite is whetted, get one of them to find out more.

Astrophotography splits into two basic varieties – long- and short-exposure. Traditional astrophotography is long-exposure work, which targets deep-sky objects and comets, and is equally feasible using either film,

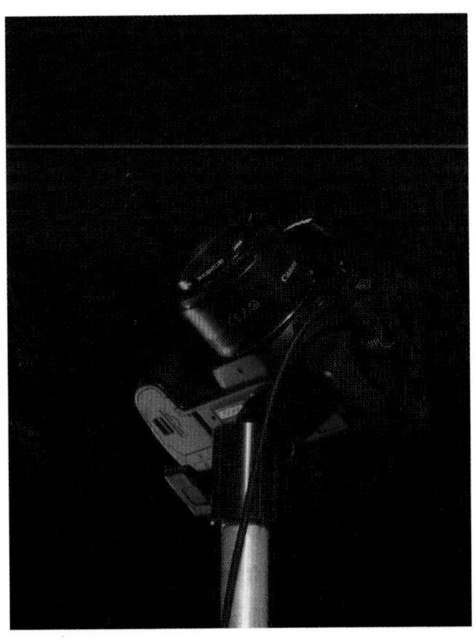

► *All you need for simple astrophotography: a digital camera with cable release on a tripod. It helps to run the camera from a separate power supply as the batteries can run down very quickly.*

SLR digital cameras or specialized cooled CCD cameras. Short-exposure photography essentially means photographing bright Solar System objects such as the Sun, Moon and bright planets and is carried out with webcams, digital cameras or CCD cameras.

Camera choice

There are many possibilities for photography of one sort or another in astronomy, and what you can achieve will depend on what astronomical equipment you have, your interests and observing conditions, and the photographic equipment you have available. So here I describe the options of what you can do with the various types of camera.

Compact film or digital cameras, restricted to short exposures (up to 1 second)

These cameras are only suitable for pointing through the eyepiece and taking a snapshot (known as the afocal method). This is far easier with digital than with film, because of the difficulty of aligning the camera with the telescope, checking the focus and getting the right exposure, which is all rather hit-or-miss.

You can buy adapters for attaching the camera to the eyepiece, either using its lens extension thread or the tripod bush, if it has one. Taking the actual shot ideally requires a cable release, in order to avoid jogging the setup as you press the shutter button. But most cheap compacts do not have this facility, in which case you must use the camera's self-timer, which then means that you can't take the shot instantaneously.

Many cheap digital cameras are restricted to a single sensitivity setting of ISO 100, which gives good color, contrast and detail, but limited sensitivity to light. Film cameras may use the full range of films available. I recommend using slide rather than print film if you rely on commercial prints because the automatic exposure control on commercial printmaking machines does not cope well with most astronomical subjects, whereas with slides you can see the results directly. However, if you scan your negatives into the

◀ An adapter that enables a digital camera to be used with the afocal method. It is fully adjustable for a wide range of different camera designs.

computer and make your own prints, this does not apply and you can use print film.

The exposure problem applies whenever the subject – say a planet – is surrounded by a large area of black sky. Auto-exposure systems, whether in the camera or in the printer, are designed to give average exposures with an equal balance of light and dark, so they increase the exposure to the extent that you just get a bright overexposed dot on a gray background. Unless you can override the camera's auto-exposure control, you are restricted to photographing the Moon or Sun, where the subject fills most of the field of view. A full-aperture filter over the telescope aperture is essential for solar work.

Compact film or digital cameras capable of longer exposures (usually up to 1 minute)

These cameras are also suitable for afocal photography of planets through a telescope, and they usually have manual exposure and focus override – features that allow you much greater control. In the case of digital cameras, there is also a choice of sensitivity settings. The longer exposure times make possible basic constellation photography – put the camera on a tripod with ISO 400 or similar setting, or film, and give a time exposure of 10 or 15 seconds.

As well as photographing constellations, this method can be used to photograph planetary alignments, bright comets, aurorae and noctilucent clouds. You might be eager to try the long-exposure capabilities of your camera to photograph deep-sky objects through the telescope, using the afocal method, but this is not easy.

ISO 400 is a good compromise between sensitivity and image quality, for both digital and film images. A practical limit with current cameras is ISO 1600, at which setting digital camera images lack detail and suffer from electronic noise, while film becomes rather grainy and low contrast. Film of ISO 100 is often called "slow," while ISO 400 is "fast," and you may hear the same terminology applied to digital cameras.

Film or digital SLR (single-lens reflex, with detachable lens), no limit to the exposure time (a B or "Bulb" setting)

The longer exposures possible with these cameras permit deeper photography of constellations using the camera's standard lens, but with the camera mounted on top of a polar-aligned driven telescope, or on the equatorial mount alone.

These SLR cameras are also ideal for photography directly through the telescope, which you can regard as a very long focal-length lens. Adapters are available for attaching all major camera brands to the telescope (see page 168).

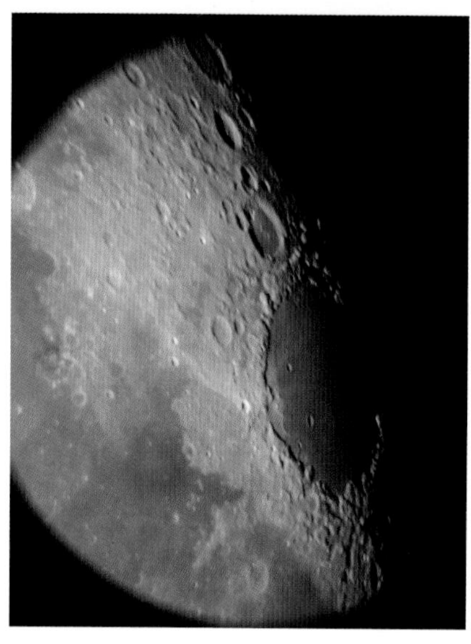

◀ *The Mare Crisium, photographed by eyepiece projection using a 215 mm Newtonian reflector by the author. The drive on this telescope is not good enough for long-exposure photography, but is quite adequate for brief lunar exposures.*

An advantage of digital SLRs is that it is usually possible to link them directly to a computer's USB port so that the images are stored on the hard drive rather than on the camera's memory card, and also to operate and control the shutter speed from the computer. This has two benefits – you can operate the camera from indoors (as long as the telescope is nearby), and you can see the images on your computer monitor as soon as they are taken and downloaded, which means that you can judge the quality of the image much better than on the camera's own LCD display.

Webcam

These cameras were developed to enable computer users to send low-resolution videos of themselves at the computer to others via the Internet, but they were quickly adapted for use at the telescope for imaging bright objects, usually those within the Solar System. They are available very cheaply, and certain brands are particularly suited to use at the telescope.

A webcam requires a computer for its operation, and generally plugs into the USB socket, which is universally available on the more recent computers. However, the computer must be within a few meters of the telescope. Webcams come with software that allows you to produce a video stream which you can then either transmit via the Internet or store as a file. The availability of webcams suitable for astro-imaging varies from time to time, so go to www.stargazing.org.uk for the latest information.

At the heart of a webcam is a light-sensitive sensor, basically the same as is found in any digital camera or video camera. It has low resolution of 320 × 240 pixels or 640 × 480 pixels in color. The main requirements for astronomical use are that you should

▲ *A single frame (left) from a 2-minute webcam sequence of Jupiter shows the typical appearance of the planet in a 200 mm reflector in average seeing. Using Registax to combine the best 900 of the 1200 frames gives the view at right after image processing. Io is visible crossing the disk at right.*

be able to remove the lens in order to attach the camera to your telescope, and that you should be able to vary the exposure parameters using the software. Removing the lens may mean dismantling the camera and making your own adapter. Having found the object you want to photograph in the eyepiece, you then replace the eyepiece by the webcam, refocus as necessary and record a short video sequence.

The other essential is specialist astronomical software to process the images for you. Many suppliers now include appropriate software, such as Registax, http://www.astronomie.be/registax/, with the camera and adapter. The software takes the hundreds or thousands of images in the sequence, chooses those that have the highest contrast (and therefore are sharpest), and stacks them together in register. It also carries out image processing to bring out the finest detail, and the results are often truly amazing.

Webcams are usually only suited to short-exposure photography, though they can be modified for longer exposures.

Video cameras

It is possible to produce similar results to webcams using commercial camcorders, though these generally have the drawback of much larger size and non-removable lenses. Some small video cameras with removable lenses are also adaptable for long-exposure imaging, and have their devotees.

Cooled CCD cameras

These cameras are most suited to long-exposure photography through the telescope, though they can equally be used for lunar or planetary photography. They incorporate electronic or liquid cooling of the CCD chip, which reduces the electronic noise that spoils many images from digital cameras. They are mostly monochrome cameras and require separate exposures through red, green and blue filters to produce a color image after image processing. At the top end, these cameras can produce very detailed and stunning images of large size, but they are expensive and require a deep involvement in the imaging process.

As with digital SLRs and webcams, you are separated from the telescope so it is possible to remain in a warm environment and to see the images as they are taken.

Practical imaging

The secret of successful astrophotography is to take it step by step. Don't go out and buy all the adapters for photography through the telescope and charge straight at it. Build up your expertise bit by bit, getting the best results at each stage. That way you'll learn the limitations of your system and your local sky, and will know straight away how to improve things. I have assumed in what follows that you are familiar with the operation of your camera and the various apertures and shutter speed settings.

Begin by using the camera just by itself, with whatever lens it comes with. At one time all cameras came with a "standard" lens of fixed focal length, usually 50 mm, but these days a zoom is more common. You will also need two other items: a tripod and a cable release. The cable release allows you to operate the shutter without jogging the camera and to keep the shutter pressed during a long exposure

◀ *Trail photos are easy to take and can be attractive: this eight-minute exposure shows the Pleiades rising behind a tree.*

by means of a screw or, more usually these days, a friction clamp that takes the form of a knurled ring at the end of the cable. Some digital SLR cameras only take an electronic cable release, which is invariably more expensive than the traditional photographic one.

Give time exposures of one, two and four minutes with the camera pointed toward the pole and a speed setting of ISO 400. The lens should be focused on infinity and its aperture should be wide open – f/1.8 or thereabouts on a standard lens, around f/3.5 on a zoom. If your skies are very dark you can give longer exposures still. The aim is to record star trails, with the rotation of the Earth moving the stars through the field of view so they leave a trail on the film. Take exposures of different parts of the sky, with varying exposure times, down to as short as 10 seconds. Include the horizon in some of the shots, concentrating on regions where there are bright stars.

The results will reveal the extent of light pollution from your site using your camera. You should have recorded some interesting patterns, and you will also get a pictorial record of the way the sky moves. In many cases you will discover one of the modern astrophotographer's enemies: aircraft. Their lights also trail across the film, spoiling many a good shot.

Toward the pole, the stars turn in arcs that get longer the farther you get from the pole itself. At the celestial equator, the trails are straight. To the east and west, stars rise and set at an angle that depends on your latitude. Notice the extent of any light pollution. From a city or suburb, even the 30-second shots will probably be strongly colored by the lighting, and the long exposures will be so overexposed that no stars are visible. From the country, sources of light pollution near the horizon will show up clearly.

Look closely at the star images themselves. Even at the celestial equator, where the stars move fastest through the sky, the 10-second exposures will be virtually points of light, whereas on the longer exposures they will have trailed. You should be able to record stars down to fainter than sixth magnitude using ISO 400, but because the Earth keeps turning, there is a limit to the faintest star you can record, no matter how long the exposure, because the light does not spend long enough on each part of the chip or film.

If your digital camera restricts you to exposure times of 30 seconds or so, it is possible to take repeated shots and combine them using software to produce the effect of star trails. This can have an advantage over film because the foreground and sky brightness of the shorter exposures can be retained, thus giving pictorial views, whereas with a single long exposure in suburban areas everything is often seriously overexposed.

Piggyback photography

To record fainter objects, the camera must follow the stars through the sky. So the next step is to mount the camera on top of your telescope somehow, and repeat the experiment with the telescope following the stars. Some telescopes make provision for mounting a camera on them piggyback, while others require a bolt-on adapter. A few have no means of mounting a camera at all, even though they may have equatorial mounts that would allow astrophotography, and this is a point to consider when choosing the telescope in the first place.

Note that telescopes on basic altazimuth mounts, such as ETXs, NexStars and many SCTs, are unsuitable for long-exposure photography because the field of view itself rotates, leading to trailing at the edge of the image, although you can add individual short exposures in image-editing software instead.

Your earlier experiments will tell you how long an exposure you can give without trouble from light pollution. If your mounting is aligned correctly on the pole and the drive is accurate, you should be able to get away with standard-lens exposures virtually as long as you want. The lucky ones in dark skies can use fast settings and long exposures with a standard lens for an hour or so and get beautiful shots of the Milky Way and constellations.

Look at the results carefully, and at the "What if..." table on page 175 to decide how to improve matters. Once you have obtained good results with a standard lens, you can progress to telephoto lenses in order to get more close-up views of large nebulae and star clusters, such as the Lagoon Nebula and the Pleiades. Such lenses are often available cheaply secondhand, particularly if your camera can take the simple screw-thread lenses, often described as Pentax screw or M42 fitting. You may be able to get an adapter for your camera to take these lenses, which could save a good deal of money. You don't need automatic lenses for astronomy – the old manual types are fine. Where lenses are concerned, "automatic" means that they have

◀ *A Canon 10D digital SLR camera mounted piggyback on the cradle of a Chinese-made Sky-Watcher 130 mm telescope with motor drive.*

a linkage that closes the aperture down automatically as you take the picture, allowing you to focus with the lens at full aperture to give a bright image. Manual 200 mm and 400 mm screw-thread lenses are available secondhand very reasonably, though the optical quality may not be assured. Make sure that the lens will focus to infinity before you buy it. Although the lens will turn to the infinity setting, it may not actually focus to infinity at that setting, which could be why it was sold off in the first place.

As the focal length increases so does the requirement for tracking the sky accurately. To make matters worse, the photographic speed, as measured by the f-number, is slower. A standard lens is usually better than f/2.0, while a 400 mm lens is around f/6.3, requiring ten times longer exposures for the same result. Drive and polar alignment errors can easily creep in, along with problems caused by vibration and wind shake.

▲ *A photograph of the Andromeda Galaxy, M31, taken with the system shown on page 166, using a 135 mm lens on the camera. The exposure time was 30 seconds with the camera working at ISO 1600.*

No drive is perfect, and trailing can be caused by periodic errors in the drive's rate as well as by poor polar alignment and mechanical defects such as sagging as the mount shifts in position – quite likely with the smaller and cheaper mounts. Some basic telescopes have a simple drive and no more – it runs at a single speed and that's it. But others allow you to correct the drive rate by pressing the buttons on a handset. One button speeds up the RA drive and the other stops it altogether, so a brief touch on either button will correct for small shifts in the image position. If there is a motor on the declination axis, another pair of buttons allows you to correct for dec errors, otherwise you will have to use the manual slow motion carefully to avoid jogging the telescope.

If you have a telescope with drive corrector buttons, it makes sense to monitor and correct the drive rate of the telescope as you take long exposures – what is called a guided, rather than a driven, exposure. For this you will need an eyepiece with crosswires, preferably illuminated. These, known as illuminated reticle eyepieces, cost about twice as much as a basic eyepiece, with those containing their own battery costing more than those with wires to a battery pack or the telescope's own power supply (and remember that the little batteries they contain are more expensive to replace than ordinary ones).

Thus equipped, you can monitor your exposures and correct for slight shifts during long exposures. And by now you are most of the way to taking shots through the telescope itself. With compact cameras you are very restricted, as already described, but webcams are very easily attached and can take excellent short-exposure images even through entry-level telescopes that would be quite inadequate for old-fashioned film photography. However, for long-exposure photography of deep-sky objects you really need either an SLR or a cooled CCD camera.

Photography through the telescope

A telescope is effectively a long telephoto lens, so you need to attach your camera to the telescope in the same way as you would a lens. You can usually buy adapters to photograph through most commercial telescopes, even the small ones. With improvements in CCDs and digital camera technology, it is now possible to take photos of deep-sky objects through instruments that would have once have been thought of as being far too small.

The standard means of adapting different camera mounts to a telescope is the T-mount system, designed in the 1950s by lens manufacturer Tamron as a means of producing lenses that would fit all cameras. You need a T-mount ring for your particular camera, plus an adapter that will attach the T-mount thread to whatever your telescope requires. This will be either a standard $1\frac{1}{4}$-inch eyepiece fitting, in the case of a telescope with a focusing mount, or a thread that attaches directly to the telescope tube, common to virtually all SCTs and the larger Russian-made Maksutovs, though unfortunately not the same as found on ETXs and NexStars, which have their own individual threads. It is well worth obtaining an adapter that allows you to incorporate an eyepiece

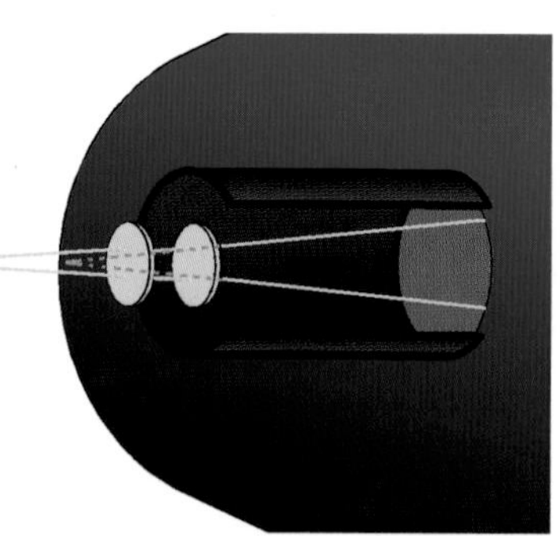

▲ *The Newtonian focus point is often very close to the focusing mount, leaving no room for a camera (dashed line). But using a Barlow lens both moves the focus point farther out and increases the effective focal length – notice that the beam converges at a different angle, as if the focal length were longer (solid line).*

► To photograph directly through the telescope requires an adapter such as this, which can hold an eyepiece inside in order to increase the effective focal length. You also need the appropriate adapter to fit your camera.

into the system, which then enables you to increase the effective focal length at which you are photographing.

The adapter system only works if you have enough focusing range on your telescope to focus the camera, so that you can take pictures at the main or prime focus point of the telescope. The focal plane of the camera is about 50 mm behind the plate where you attach the lens, so you will need to bring the camera this distance closer to the main mirror. Most standard SCTs, Maksutovs and refractors, apart from the smallest, should have this focusing range. Some Newtonian focusers, however, do not, in which case you will need to use an eyepiece in the adapter at all times to extend the focus point. This also increases the effective focal length by a factor that depends on the eyepiece and its position in front of the camera's focal plane. You can get adapters that allow you to vary the distance between the eyepiece and the camera, giving you greater control over the effective focal length.

Whatever telescope and camera system you use, if you want to photograph the Moon and planets, increasing the effective focal length is necessary in many cases where you want a magnified image. The image of Jupiter, say, at the prime focus of a 2000 mm focal length telescope is only half a millimeter across and Mars is only half that size, even when at its closest. Just focusing the image using an SLR is a challenge when faced with such small images, and is even more difficult where all you can see are stars. Some digital SLR cameras now have "live view," which means that you can check the focus in the viewfinder in real time, or better still on a computer linked to the camera.

For some objects, however, the opposite applies. The f/10 focal ratio of a standard SCT is rather long and slow for photographing some objects, such as the larger galaxies and nebulae, in which case you can buy a focal reducer or telecompressor, which changes the f/10 to a faster f/6.3. In some units the optics also flatten the field, avoiding out-of-focus stars at the field edge. Some of these accessories can also be used for giving wider-field visual views, though the sky brightness is also increased.

Another similar device is the coma corrector, which may be needed on short-focus Newtonian and Dobsonian telescopes. The best images from any telescope are right on its optical axis, and as you get away from this axis the image of a star tends to elongate to a teardrop shape. For most of the time this is unnoticeable, but it can be objectionable when you try to get wide-field views from a short-focus reflector. Visually this can be caused by the eyepiece being unable to handle off-axis images, but if it persists when taking pictures at prime focus a coma corrector should help. This fits between the telescope and the camera and can also be used visually. But the cost, as you might expect, is high – around the same as a super-wide-field eyepiece.

Having attached your camera to the telescope, you can now experiment with photographing objects using brief exposures, which effectively means the Moon and bright planets. Exposure times vary considerably depending on the instrument, the effective focal length, the film and the conditions, but as a guide to exposure times, give a fraction of a second exposure for the Moon and up to a second or two for the brighter planets. Having said that, virtually all planetary and lunar photography is now done using webcams or CCDs, which can give large numbers of very brief exposures. However, to photograph fainter objects, such as nebulae and galaxies, you will need to give much longer exposures, and this is where cooled CCDs in particular come into their own. Half a minute is adequate for showing the brighter deep-sky objects, but several minutes to hours are needed for faint objects.

Unless you are very fortunate, such long exposures will show up all sorts of defects in the drive, alignment and guiding of the telescope. But how can you monitor these when the camera is attached to the telescope and all its light is falling on the film or chip? There are two main methods. One involves using a separate telescope attached to the main one specifically so that you can monitor the drive rate. This telescope should have a focal length at least half that of the main telescope, and must be very securely mounted on it to avoid flexure between the two.

◀ *Celestron's Radial Guider provides a means of monitoring the drive rate during a long exposure by using a prism to view a star just outside the main field of view of your image. Most people now use a CCD autoguider to watch the star and make automatic corrections.*

► *The reflection nebula NGC 1977, near the Orion Nebula, photographed by Philip Perkins using a 250 mm Meade SCT with a Lumicon off-axis guider and an ST-4 autoguider. The total exposure time was over three hours. He calls this the "Ghost Nebula" after the shadowy figure in the center.*

The other method is to use an off-axis guider attachment. This consists of a tube that fits between the main telescope and the camera, within which is a small prism that collects light from the part of the image that would be just outside the frame area of the camera, some way from the main optical axis of the telescope. You must choose a guide star just outside the field of view of the object being photographed, which is not always easy. At one time you would spend the duration of the exposure watching the guide star's every movement, and correcting for its shifts in position. This called for superhuman efforts on a cold night and when using film you couldn't always be sure that the results would be worth it. But now most people would use an autoguiding system (see page 173), either with the off-axis guider or the separate guide scope. There are plenty of things to go wrong in astrophotography without the effort of guiding for long periods.

CCD imaging

A CCD is a silicon chip consisting of an array of light-sensitive elements or pixels. The image can be collected over a period of time and read out into a computer that constructs the image. One of the great advantages is that the image can be further processed to bring out nuances of detail that might otherwise be lost in a photograph. Astronomers rapidly took to the great sensitivity to light and versatility of CCDs, and professionals now use nothing else. Although every digital camera and webcam uses a version of a CCD, there is a great advantage in using a purpose-built cooled CCD camera because of the much lower electronic noise in the image.

While you can carry out digital photography using just the camera,

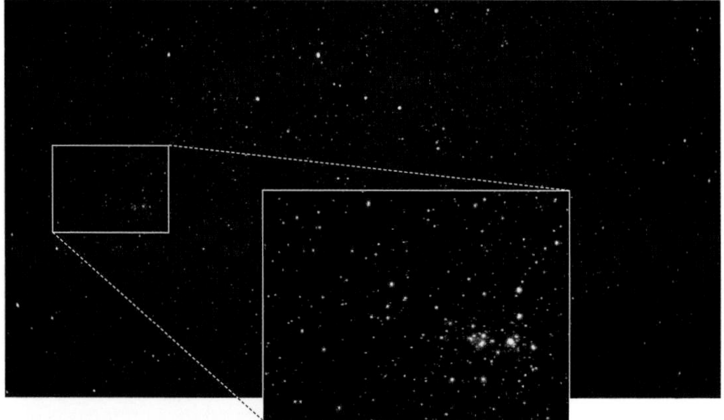

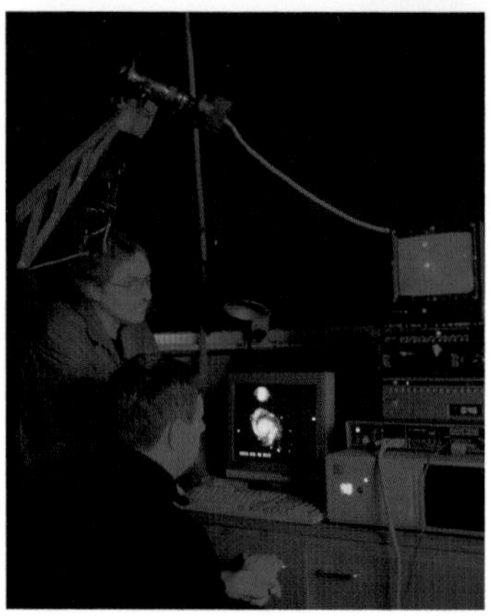

▲ *The main photo shows the approximate area of Cassiopeia covered by a standard 50 mm lens on 35 mm film. The small rectangle shows the area covered by a typical CCD using the same lens, in this case the Starlight Xpress MX5-C color camera, with the image it produces shown in the larger inset.*

◀ *Using a CCD on a 450 mm reflector, members of Salford Astronomical Society in the UK can record images of galaxies that are barely visible from their light-polluted observatory.*

with CCDs as with webcams you do need to have the computer fairly close to the telescope. For many people this means using a laptop, with the attendant problems of power supply.

The currently available CCD cameras have CCD arrays that are mostly smaller than the traditional 35 mm film frame. Your aim therefore needs to be spot-on simply to get the object within the imaging area. Even the more expensive CCD chips are still only a fraction of the size of the 24 × 36 mm film frame.

Color CCD imaging usually means taking separate exposures through red, green and blue filters and combining the results afterward during image processing, though single-shot color CCD cameras are also available.

All this might make it sound as if CCD imaging is more trouble than it is worth. But there is one advantage that tips the balance for many enthusiasts. Whereas film gives very low contrast at the low-exposure end of the scale, making it very difficult to pick out an object from the background, the response of a CCD to light is linear. With image processing, it's possible to subtract the background – including light pollution – to leave only the object. This property makes it possible to take remarkable images of galaxies and other deep-sky objects from light-polluted urban skies even though the objects themselves are completely invisible in any telescope from that site.

You could argue that better images can be taken from dark sites, but there is a great appeal in seeing the image of a spiral galaxy appear on your monitor despite hardly being able to see the brightest stars visually. It is a matter of personal achievement, as worthwhile as any other. Just as a sports player is not content to leave the game up to the professionals on the grounds that they are so much better at it, so it is in amateur astronomy.

Advanced gadgets

Soon after the introduction of the CCD came the autoguider. This is a unit based on a simple CCD that detects the direction of movement of a star image and automatically operates the drive correction motors to bring the star back to the center. Astrophotographers who were fed up with guiding their telescopes for long periods during exposures welcomed autoguiders with open arms. The standard method is to use the autoguider with an off-axis guider. It will only work with fully driven telescopes. Many people use cheap webcams, though these require fairly bright guide stars. Some CCD units have autoguiders incorporated, for easier CCD imaging.

Another clever device to improve your image is the adaptive optics unit. This constantly monitors the state of the seeing by observing a star near the object you are imaging or photo-

▶ *The Meade Deep Sky Imager, seen here on a Sky-Watcher 80 mm refractor, is an uncooled CCD unit. Software stacks short exposures in registration to give the effect of a single long exposure.*

graphing, and constantly corrects the position of the whole image by moving a glass element in the optical path.

Owners of Dobsonian or altazimuth telescopes need not feel left out of the computer revolution. Add-on encoders for the axes of most telescopes are available that will sense the telescope's position, once correctly set up. The units come with a handset loaded with a database of objects, just like those of commercial computer-controlled instruments. But as there is no motor, the handset displays symbols that change as you move the telescope, so that you can move it by hand to the chosen object.

More computer wizardry

Amateur astronomers have taken to computers in a big way. In fact, it is rare to find an active observer these days who does not use one for planning or analysing observations, possibly directing the telescope, and communicating with the rest of the community using e-mail and the Internet.

Computer-based sky maps are among the most valuable telescope accessories. They range from those with more popular appeal, such as *Starry Night*, to those designed specifically for the amateur

▼ The Rosette Nebula in Monoceros, photographed with a CCD camera by David Arditti using an 80 mm ED refractor from light-polluted Edgware, in the London suburbs. He used a total of over three hours of short exposures through narrowband filters, including the green O III filter.

astronomer, such as *SkyMap*. They display the sky at any chosen scale and with the user's choice of limiting magnitude and objects displayed. Most use the *Hubble Guide Star Catalog*, which includes stars down to 15th magnitude, although the accuracy of the magnitudes displayed is not particularly good at the fainter end. This catalog is only available on CD-ROM versions of software. A number of professional catalogs of double stars, clusters, nebulae and galaxies are also included, as well as the positions of many Solar System bodies such as the Sun, Moon and planets, and even some asteroids and comets.

Such programs will show the sky from your exact location at any time, including the Moon and planets, which of course an ordinary star map will not do. This makes it easier to plan your night's observing – you can tell at a glance exactly when the Moon will pop its head above the horizon, or a planet is highest in the sky, for example. You can print out finder charts for faint objects for use at the eyepiece, and identify objects you have spotted. Many of them also contain drivers that enable you to link them to a telescope that already has computer-control options. This means that you can click on an object or part of the sky and the telescope will find it for you.

What if . . .

. . . a driven photograph shows short, straight trails?
Either the rate of the telescope's drive motors is faulty or there is slippage somewhere in the system. If the trailing is in the direction of RA, then the motor is running at the wrong rate. Make sure that the axis clamps are tight so that there is no slippage between the drive and the axis, and that all other screws are tight. If all else fails, is the mount on firm ground? A tripod leg may slowly sink into damp soil over the period of the exposure.

. . . the photograph shows circular trails, but not centered on the pole?
In this case, the polar alignment is wrong, or you are trying to use an altazimuth mounting.

. . . stars at the center of the field are sharp but those at the edge are defocused or have odd shapes?
The lens is not perfect. Close the aperture down by at least one stop (f-number).

. . . stars appear misty?
Dew has formed on the lens during the exposure.

. . . there is not enough detail in the picture – all the images appear grainy?
Fast film is notorious for its grainy results. Use slower film and give longer exposures. On digital cameras, use a lower ISO setting.

. . . all I get is a white or pale yellow image?
The light pollution at your site is so strong that it has wiped out all the stars. Either use shorter exposures or a slower ISO setting, or, ideally, use a darker site. Digital camera exposures can be adjusted using the "Levels" control of image-editing software such as Photoshop.

Appendix One
Using the sky maps

These maps divide the whole sky into six sections. The north and south circumpolar regions are on this spread, while the middle regions of the sky are on the following two spreads.

To see which stars are visible at any time, first look at the maps of the middle regions of the sky. Choose the map that corresponds to the month in which you are observing, and if you wish allow for the exact date and time, as described on page 108.

The stars on that map are now visible on your meridian, looking south in the northern hemisphere and looking north in the southern. Use the maps the right way up in the northern hemisphere and upside down in the southern. Stars at the bottom of each map from your point of view may be below your horizon, while stars at the top will be roughly overhead.

Northern polar regions

The northern and southern polar regions are visible in each hemisphere looking north and south respectively. The appropriate month is shown at the top of each map. Again, stars at the bottom of each map may be below your horizon while stars at the top are roughly overhead. All the maps overlap each other so you can link from each one to the adjacent part of the sky.

Stars are shown down to magnitude 5, which is often the magnitude limit in outer suburban areas. Only the brightest stars are named. Selected deep-sky objects are shown on each map, mostly those visible with binoculars. Messier objects are shown as, for example, M31, while NGC objects are shown by a number only, NGC 752 being shown as 752.

The red line crossing the maps is the ecliptic, while the wide pale band is the Milky Way. Declinations and right ascensions are shown in gray. Grey dotted lines indicate the borders of the constellations.

These maps were produced with the help of the *SkyMap* computer program.

Southern polar regions

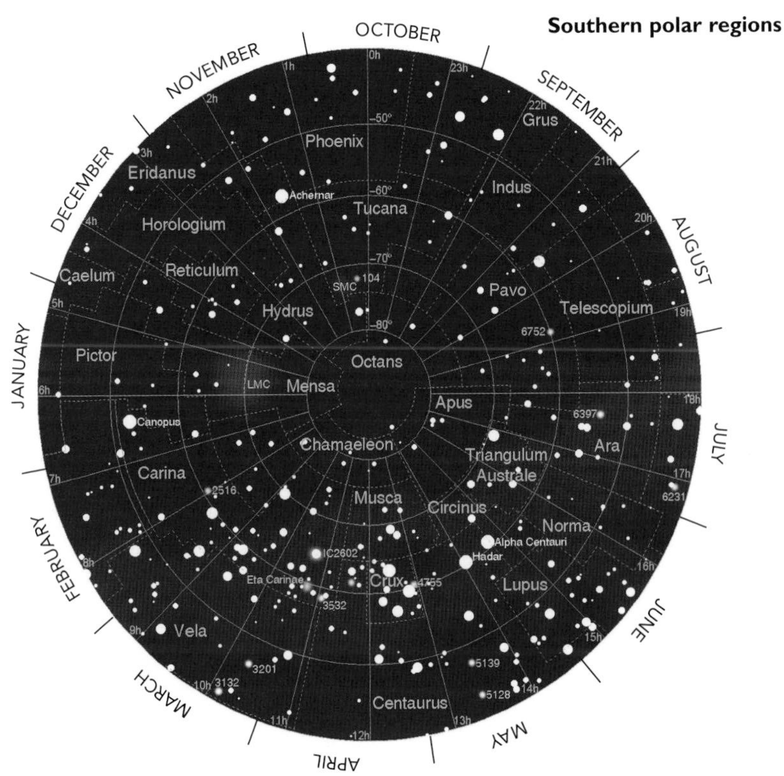

Equatorial regions: September to November (RA 21h to 3h)

This part of the sky contains the point where the Sun, on the ecliptic, crosses the celestial equator moving northward, but it is an area almost completely devoid of bright stars. Apart from the splendid Andromeda Galaxy, M31, visible with the naked eye, there are few major deep-sky objects to view at this time of year.

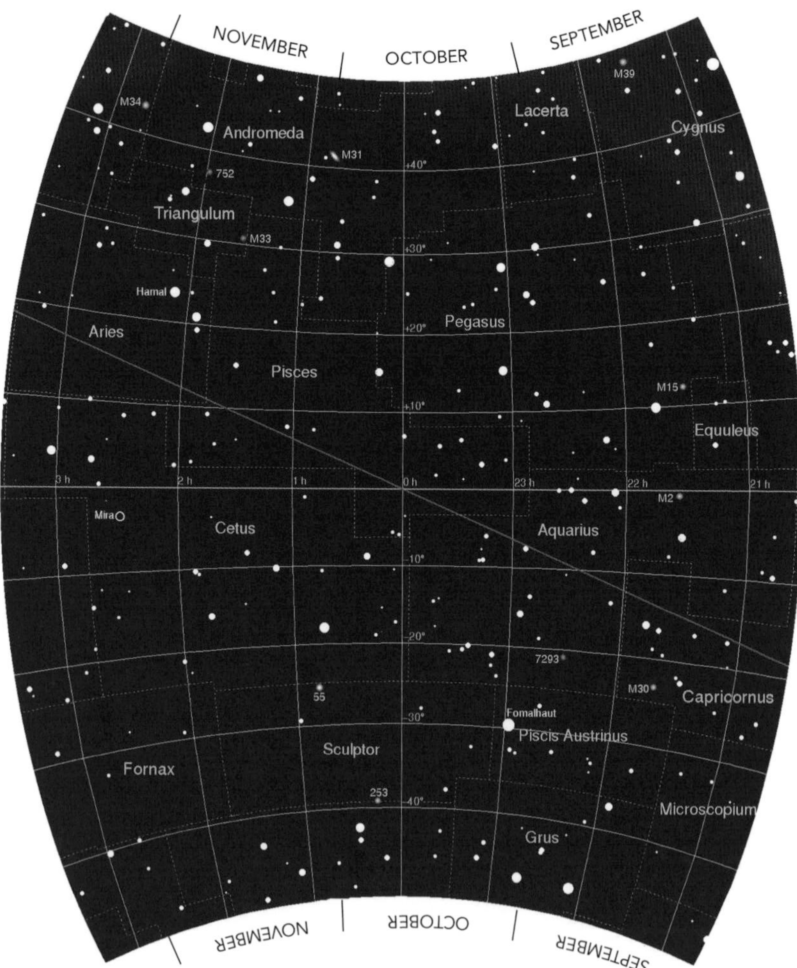

Equatorial regions: June to August (RA 15h to 21h)

The Milky Way is at its best at this time of year, the stretch from Cygnus to Sagittarius being the brightest. Bright nebulae and open clusters are scattered down its length, and whether you are observing with the naked eye, binoculars or a telescope there is plenty to see. The dark clouds of the Great Rift are also obvious to the naked eye.

Equatorial regions: March to May (RA 9h to 15h)

By chance, this region of the sky is well away from the obscuring dust clouds of the Milky Way, allowing us to view the huge Virgo cluster of galaxies which lie in this direction. Don't expect a dramatic sight – none of them are visible with the naked eye and they are inconspicuous with binoculars or small telescopes. But this is a happy hunting ground for deep-sky observers.

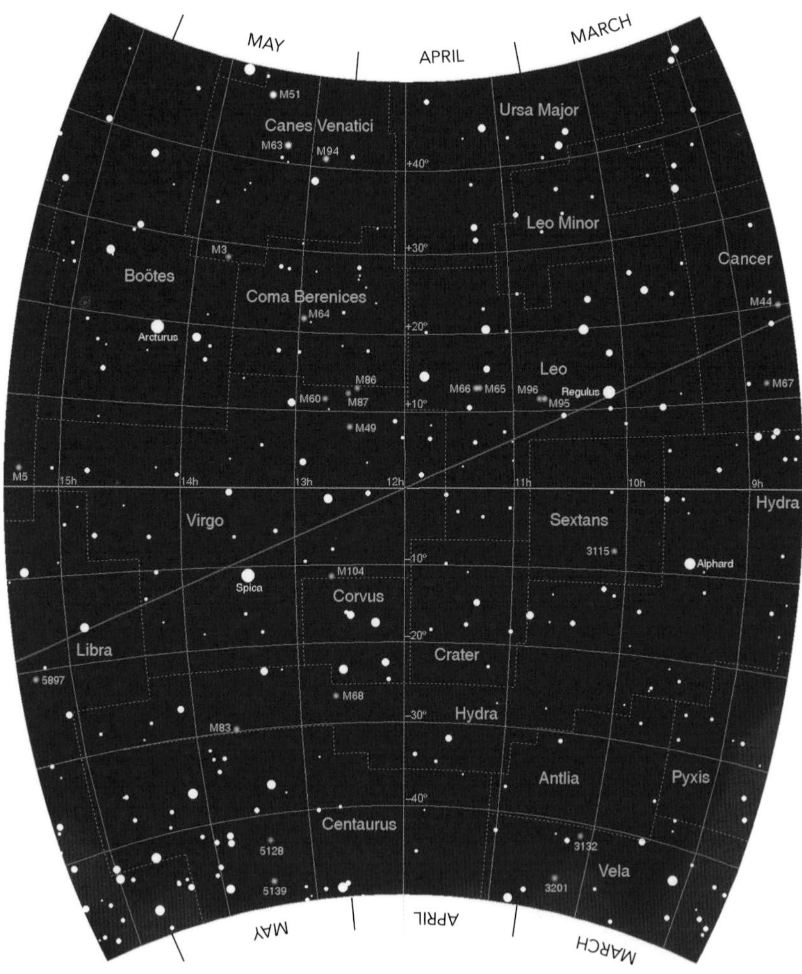

Equatorial regions: December to February (RA 3h to 9h)

Although the Milky Way at this time of year is not prominent, the stars in the nearby Orion spiral arm make up for it. Orion itself forms an ideal signpost to the other constellations. Nebulae and clusters dominate, in particular the famous Orion Nebula, M42, which is visible with the naked eye and a dramatic sight using any optical aid.

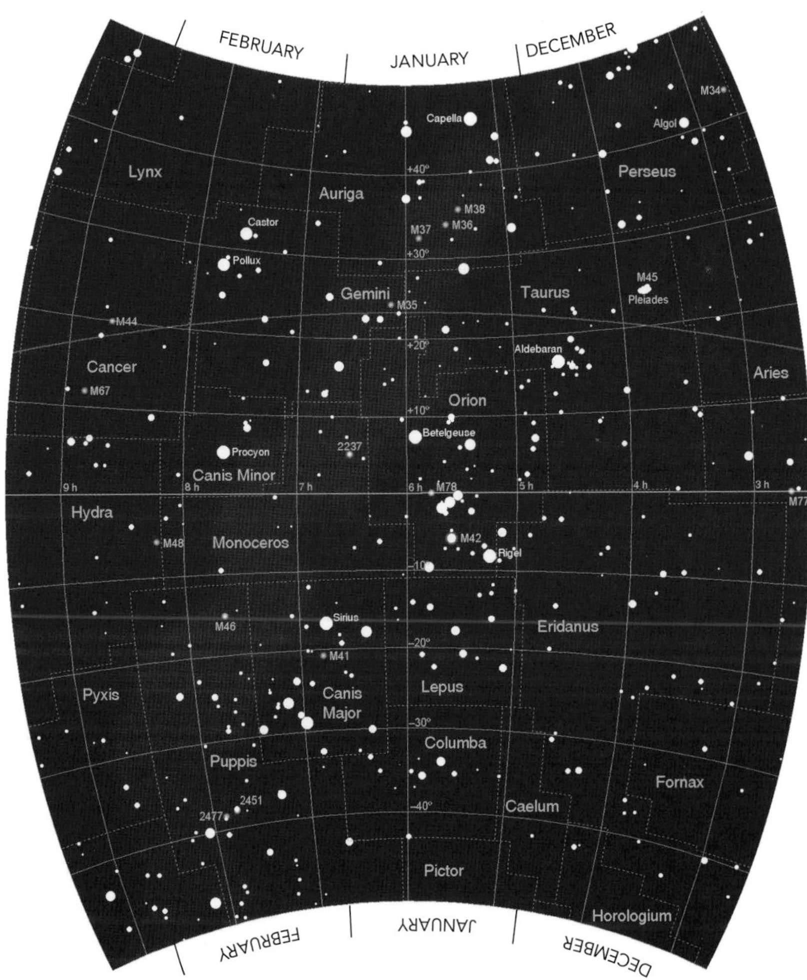

Appendix Two
Interesting objects to observe

This is a list to help you find a range of deep-sky objects with telescopes of all sizes, listed by object type. Each list is in order of declination, so northern-hemisphere observers will find that the objects at the top of the list are the highest in their sky, and vice versa in the southern hemisphere.

The magnitude and size of objects are a guide as to what to look for, but they are only very approximate. If the total brightness is spread over a large area, it could be hard to see, whereas a small total brightness concentrated in a small area can be surprisingly easy. The Comments column gives any popular name for the object, along with an idea of the smallest instrument size, in millimeters, you would need to see it comfortably under fairly dark, but not necessarily ideal, conditions. This is only a guide and should not be taken too literally; light-pollution filters may be needed. Bins is an abbreviation for binoculars and NE means naked eye.

GALAXIES									
Object	Constellation	RA		Dec		Mag	Size	Type	Comments
		h	m	°	′		arc min		
M82	Ursa Major	09	55.8	+69	41	8.4	11 × 5	EOS	Binoculars
M81	Ursa Major	09	55.6	+69	04	6.8	26 × 14	S	Binoculars
NGC 2403	Camelopardalis	07	36.9	+65	36	8.4	18 × 10	FOS	60
NGC 5907	Draco	15	15.9	+56	20	10.4	12.3 × 1.8	EOS	200
M108	Ursa Major	11	11.5	+55	40	10.0	8 × 2	EOS	75
M101	Ursa Major	14	03.2	+54	21	7.7	27 × 26	FOS	Binoculars
NGC 2841	Ursa Major	09	22.0	+50	58	9.3	8.1 × 3.8	S	60
NGC 147	Cassiopeia	00	33.2	+48	30	9.3	13 × 8	E	300
NGC 185	Cassiopeia	00	39.0	+48	20	9.2	12 × 10	E	60
M106	Canes Venatici	12	19.0	+47	18	8.3	18 × 8	S	50
M51	Canes Venatici	13	29.9	+47	12	8.1	11 × 8	FOS	Bins; Whirlpool
NGC 4449	Canes Venatici	12	28.2	+44	06	9.4	5 × 4	I	60; Box Galaxy
NGC 891	Andromeda	02	22.6	+42	21	9.9	14 × 3	EOS	100
M63	Canes Venatici	13	15.8	+42	02	8.6	12 × 8	S	60; Sunflower
M31	Andromeda	00	42.7	+41	16	3.5	180 × 63	S	NE; Andromeda Galaxy
M94	Canes Venatici	12	50.9	+41	07	8.1	11 × 9	S	60
M32	Andromeda	00	42.7	+40	52	8.2	8 × 6	E	60
NGC 404	Andromeda	01	09.4	+35	43	10.4	4.4 × 3.3	S	100
NGC 7331	Pegasus	22	37.1	+34	25	9.5	11 × 4	S	50
NGC 4656	Canes Venatici	12	44.0	+32	10	10.6	15 × 3	P	100; Fish Hook
M33	Triangulum	01	33.9	+30	39	5.7	62 × 39	FOS	Bins; Pinwheel
NGC 3344	Leo Minor	10	43.0	+24	54	10.5	6 × 5.1	FOS	60
M64	Coma Berenices	12	56.7	+21	41	8.5	9 × 5	S	60; Black Eye
NGC 2903	Leo	09	32.2	+21	30	8.9	13 × 6	S	60

Object	Constellation	RA h	RA m	Dec °	Dec ′	Mag	Size arc min	Type	Comments
NGC 7814	Pegasus	00	03.3	+16	09	10.5	6 × 2	S	100
M100	Coma Berenices	12	22.9	+15	49	9.4	7 × 6	FOS	Binoculars
M74	Pisces	01	36.7	+15	47	9.2	10 × 9	FOS	100
M98	Coma Berenices	12	13.8	+14	54	10.1	10 × 3	S	60
M99	Coma Berenices	12	18.8	+14	25	9.8	5	FOS	60
NGC 3628	Leo	11	20.3	+13	36	9.5	14.8 × 3.8	EOS	150
NGC 4216	Virgo	12	15.9	+13	09	10	8.3 × 2.2	EOS	125
M65	Leo	11	18.9	+13	05	9.3	10 × 3	S	60
M66	Leo	11	20.2	+12	59	9.0	9 × 4	S	60
M86	Virgo	12	26.2	+12	57	9.2	7 × 5	E	60
M105	Leo	10	47.8	+12	35	9.3	5 × 4	E	60
M84	Virgo	12	25.1	+12	53	9.3	5 × 4	E	60
M87	Virgo	12	30.8	+12	24	8.6	7	E	60
M96	Leo	10	46.8	+11	49	9.2	7 × 5	S	60
M58	Virgo	12	37.7	+11	49	9.8	5 × 4	S	60
M95	Leo	10	44.0	+11	42	9.7	7 × 5	S	60
M59	Virgo	12	42.0	+11	39	9.8	5 × 3	E	60
M60	Virgo	12	43.7	+11	33	8.8	7 × 6	E	60
NGC 5248	Bootes	13	37.5	+08	53	10.2	6 × 4	S	150
M49	Virgo	12	29.8	+08	00	8.4	9 × 7	E	50
M61	Virgo	12	21.9	+04	28	9.7	6 × 5	FOS	75
M77	Cetus	02	42.7	−00	01	8.8	7 × 6	S	60
NGC 3115	Sextans	10	05.2	−07	43	9.1	8 × 3	EOS	75; Spindle Galaxy
M104	Virgo	12	40.0	−11	37	8.3	9 × 4	E	100; Sombrero
NGC 6822	Sagittarius	19	44.9	−14	48	8.8	20 × 10	I	150; Barnard's Dwarf
NGC 1300	Eridanus	03	19.8	−19	24	11.1	6 × 3.2	BS	150
NGC 247	Cetus	00	47.2	−20	45	9.5	18 × 5	S	100
NGC 253	Sculptor	00	47.6	−25	17	7.1	25 × 7	EOS	Binoculars
M83	Hydra	13	37.0	−29	52	8.2	11 × 10	FOS	75
NGC 1097	Fornax	02	46.3	−30	17	9.2	9 × 7	BS	75
NGC 2997	Antlia	09	45.6	−31	11	10.6	8.1 × 6.5	FOS	200
NGC 1365	Fornax	03	33.7	−36	08	9.5	8 × 3.5	BS	75
NGC 5102	Centaurus	13	21.9	−36	39	9.6	9.3 × 3.5	S	75
NGC 55	Sculptor	00	15.1	−39	13	8.2	32 × 6	EOS	60
NGC 5128	Centaurus	13	25.5	−43	01	7.0	18 × 14	P	Bins; Centaurus A
NGC 4945	Centaurus	13	05.4	−49	28	8.7	20 × 4	BS	60
LMC	Dorado	05	23.6	−69	45	0.1	650 × 550	I	NE; Large Magellanic Cloud
SMC	Tucana	00	52.7	−72	50	2.3	280 × 160	I	NE; Small Magellanic Cloud

BS = barred spiral; C = colliding galaxies; E = elliptical; FOS = face-on spiral; EOS = edge-on spiral; I = irregular; P = peculiar; S = spiral.

GLOBULAR CLUSTERS								
Object	Constellation	RA h	m	Dec °	′	Mag	Size arc min	Comments
M92	Hercules	17	17.1	+43	08	6.5	11.0	Binoculars
NGC 2419	Lynx	07	38.1	+38	53	10.3	4.1	250
M13	Hercules	16	41.7	+36	28	5.9	17.0	Fine binocular object
M56	Lyra	19	16.6	+30	11	8.2	7.0	60
M3	Canes Venatici	13	42.2	+28	23	6.4	16.2	Binoculars
M71	Sagitta	19	53.8	+18	47	8.3	7.2	60
M53	Coma Berenices	13	12.9	+18	10	7.6	12.6	60
NGC 5053	Coma Berenices	13	16.4	+17	42	9.9	10.5	150
NGC 7006	Delphinus	21	01.5	+16	11	10.5	2.8	200; very distant
M15	Pegasus	21	30.0	+12	10	6.4	12.0	50
NGC 6934	Delphinus	20	34.2	+07	24	8.9	3.2	75
M5	Serpens Cauda	15	18.6	+02	05	5.8	17.0	Binoculars
M2	Aquarius	21	33.5	−00	49	6.5	13	Binoculars
M12	Ophiuchus	16	47.2	−01	57	6.6	14	Binoculars
M14	Ophiuchus	17	37.6	−03	15	7.6	12	50
M10	Ophiuchus	16	57.1	−04	06	6.6	15	50
NGC 6712	Scutum	18	53.1	−08	42	8.2	7.2	75
NGC 5897	Libra	15	17.4	−21	01	8.9	12.6	200
M80	Scorpius	16	17.0	−22	59	7.2	9.0	60
M30	Capricornus	21	40.4	−23	11	7.5	11	Binoculars
M22	Sagittarius	18	36.4	−23	54	5.1	24	NE
M79	Lepus	05	24.4	−24	32	8.0	9.0	50
M28	Sagittarius	18	24.5	−24	52	7.0	11	Binoculars
M19	Ophiuchus	17	02.6	−26	16	7.2	14	50
M4	Scorpius	16	23.6	−26	32	5.9	26	Binoculars
M68	Hydra	12	39.5	−26	45	8.2	12	60
NGC 6304	Ophiuchus	17	14.5	−29	28	8.4	6.8	75
M55	Sagittarius	19	40.0	−30	58	7.0	19	Binoculars
NGC 6441	Scorpius	17	50.2	−37	03	8.0	3.0	75
NGC 5986	Lupus	15	46.1	−37	47	7.1	9.8	60
NGC 1851	Columba	05	14.1	−40	03	7.3	11	50
NGC 6541	Corona Australis	18	08.0	−43	42	6.6	13	Binoculars
NGC 6388	Scorpius	17	36.3	−44	45	6.8	4.0	60
NGC 3201	Vela	10	17.6	−46	25	6.7	18	Binoculars; large and loose
NGC 5139	Centaurus	13	26.8	−47	29	3.7	36	NE; Omega Centauri
NGC 6532	Ara	17	25.5	−48	25	8.1	7.0	75
NGC 5286	Centaurus	13	46.4	−51	22	7.6	9.0	50
NGC 1261	Horologium	03	12.3	−55	13	8.4	7.0	75
NGC 6397	Ara	17	40.7	−53	40	5.6	26	Bins; only 8400 l.y. away
NGC 6752	Pavo	19	10.9	−59	59	5.4	20	Binoculars; fine object
NGC 2808	Carina	09	12.0	−64	52	6.3	13.8	Binoculars
NGC 362	Tucana	01	03.2	−70	51	6.6	13	Binoculars
NGC 4833	Musca	12	59.6	−70	53	7.3	14	50
NGC 104	Tucana	00	24.1	−72	05	4.0	31	Naked eye; 47 Tucanae
NGC 4372	Musca	12	25.8	−72	40	7.8	19	50

PLANETARY NEBULAE								
Object	Constellation	RA h	m	Dec °	′	Mag	Size arc sec	Comments
NGC 40	Cepheus	00	13.0	+72	32	12.4	36	100; two bright arcs
NGC 6543	Draco	17	58.6	+66	38	8.1	23 × 17	60; Cat's Eye Nebula
NGC 1501	Camelopardalis	04	07.0	+60	55	12.0	52	150; sharp-edged shell
M97	Ursa Major	11	14.8	+55	01	9.9	180	60; look out for two "eyes"
M76	Perseus	01	42.2	+51	34	11.0	120 × 60	60; Little Dumbbell
NGC 6826	Cygnus	19	44.8	+50	31	8.8	27 × 24	60; blinking planetary
NGC 7048	Cygnus	21	14.2	+46	16	11.3	60 × 50	150; doughnut shaped
NGC 7662	Andromeda	23	25.9	+42	33	9.0	30	60; Blue Snowball
M57	Lyra	18	53.6	+33	02	8.8	86 × 62	50; Ring Nebula
NGC 1514	Taurus	04	09.0	+30	47	10.0	120 × 90	60; broken ring shape
NGC 6210	Hercules	16	44.5	+23	48	10.0	20 × 13	125; bright bluish disc
M27	Vulpecula	19	59.6	+22	43	7.3	480 × 340	Bins; Dumbbell Nebula
NGC 6905	Delphinus	20	22.2	+20	60	12.0	44 × 38	75
NGC 2392	Gemini	07	29.2	+20	55	9.2	43 × 47	60; Eskimo; vivid blue
Abell 21	Gemini	07	29.0	+13	15	10.2	600 × 360	200; Medusa Nebula
NGC 2022	Orion	05	42.1	+09	05	12.3	28 × 27	100
NGC 6572	Ophiuchus	18	12.1	+06	51	9.0	15 × 12	100
NGC 7009	Aquarius	21	04.2	−11	22	8.0	28 × 23	150; Saturn Nebula
NGC 1535	Eridanus	04	14.2	−12	44	9.0	20 × 7	100
NGC 2440	Puppis	07	41.9	−18	13	10.8	74 × 32	60; bluish ellipse
NGC 3242	Hydra	10	24.8	−18	38	7.8	45 × 36	150; Ghost of Jupiter
NGC 4361	Corvus	12	24.5	−18	48	10.5	80	150
NGC 6445	Sagittarius	17	49.3	−20	01	11.2	35 × 30	150
NGC 7293	Aquarius	22	29.6	−20	48	7.3	720 × 960	Helix; brightest planetary but hard because of its size
NGC 6369	Ophiuchus	17	29.2	−23	46	11.0	30 × 29	125; Little Ghost
NGC 1360	Fornax	03	33.3	−25	51	9.4	500 × 300	Best in binoculars or RFTs
NGC 6302	Scorpius	17	13.7	−37	06	9.6	83 × 24	100; Bug Nebula
NGC 6337	Scorpius	17	22.3	−38	29	12.5	48	200; mini ring nebula
NGC 3132	Vela	10	07.0	−40	26	8.5	90 × 47	150; Eight Burst Nebula
IC 4406	Lupus	14	22.4	−44	09	10.6	100 × 30	150; box shaped
NGC 3195	Chameleon	10	09.5	−80	52	12.0	38	250

PHOTOGRAPHIC OBJECTS							
Object	Constellation	RA h	m	Dec °	′	Size arc min	Comments
NGC 7822	Cepheus	00	01.6	+68	32	90 × 20	Near Cederblad 214 nebula
IC 1795	Cassiopeia	02	26.6	+62	03	27 × 13	Next to IC 1805
IC 1805	Cassiopeia	02	35.5	+61	04	90	Heart Nebula
IC 1396	Cepheus	21	39.1	+57	28	170 × 130	
NGC 281	Cassiopeia	00	53.2	+56	37	35 × 30	Pacman Nebula
NGC 7000	Cygnus	20	58.8	+44	20	120	North America Nebula
	Cygnus	20	25.2	+40	20	100	Gamma Cygni region
NGC 1499	Perseus	04	03.3	+36	25	145 × 40	California Nebula
S 147	Taurus	05	39.0	+28	00	120	Supernova remnant

Object	Constellation	RA h	RA m	Dec °	Dec ′	Size arc min	Comments
NGC 1435	Taurus	03	47.0	+23	46	30 × 20	Merope Nebula in Pleiades
NGC 2237-9	Monoceros	06	32.3	+05	03	80 × 60	Rosette Nebula
Sh 2-276	Orion	05	42.0	−01	00	400 × 20	Barnard's Loop
B 33	Orion	05	40.9	−02	28	6 × 4	Horsehead Nebula
IC 2118	Eridanus	05	04.5	−07	12	160 × 60	Witch's Head Nebula
IC 2177	Canis Major	07	04.9	−11	14	120 × 40	Seagull Nebula
	Ophiuchus	16	55.0	−23	00	240	Rho Ophiuchi Complex
NGC 6559	Sagittarius	18	10.0	−24	08	9 × 8	Near IC 1274
NGC 6334	Scorpius	17	20.3	−35	49	40 × 25	Cat's Paw Nebula
IC 4628	Scorpius	16	57.0	−40	20	40 × 25	
	Vela	08	40.0	−44	00	6 deg	Vela supernova remnant

OPEN CLUSTERS

Object	Constellation	RA h	RA m	Dec °	Dec ′	Mag	Size arc min	Comments
NGC 188	Cepheus	00	44.4	+85	20	8.1	14	Binoculars
M52	Cassiopeia	23	24.2	+61	35	6.9	13	60
NGC 663	Cassiopeia	01	46.0	+61	15	7.1	16	Binoculars
M103	Cassiopeia	01	33.2	+60	42	7.0	6	Binoculars
NGC 7510	Cepheus	23	11.5	+60	34	7.9	4	100
NGC 457	Cassiopeia	01	19.1	+58	20	6.4	13	Binoculars; Owl Cluster
NGC 869/884	Perseus	02	19.0	+57	09	4.3	30	NE; Double Cluster
		02	22.4	+57	07	6.1	30	
NGC 7789	Cassiopeia	23	57.0	+56	44	6.7	16	60
NGC 7243	Lacerta	22	15.3	+49	53	6.4	21	Binoculars
M39	Cygnus	21	32.2	+48	26	4.6	32	Naked eye
M34	Perseus	02	42.0	+42	47	5.2	35	Binoculars
NGC 2281	Auriga	06	49.3	+41	04	5.4	15	Bins; Broken Heart Cluster
NGC 752	Andromeda	01	57.8	+37	41	5.7	50	Binoculars
M38	Auriga	05	28.7	+35	50	6.4	21	Binoculars
M36	Auriga	05	36.1	+34	08	6.0	12	Binoculars
M37	Auriga	05	52.4	+32	33	5.6	24	Naked eye
M35	Gemini	06	08.9	+24	20	5.1	28	Naked eye
M45	Taurus	03	47.0	+24	07	1.2	110	NE; Pleiades
M44	Cancer	08	40.1	+19	59	3.1	95	NE; Beehive or Praesepe
NGC 1647	Taurus	04	46.0	+19	04	6.4	45	Binoculars
M67	Cancer	08	50.4	+11	49	6.9	30	Binoculars
NGC 6633	Ophiuchus	18	28.0	+06	34	4.6	20	Binoculars
NGC 2244	Monoceros	06	32.4	+04	52	4.8	24	Bins; within Rosette Nebula
NGC 2301	Monoceros	06	51.8	+00	28	6.0	12	50
NGC 1981	Orion	05	35.0	−04	26	4.6	25	Binoculars
M48	Hydra	08	13.8	−05	48	5.8	54	Binoculars
M11	Scutum	08	51.1	−06	16	5.8	14	Bins; Wild Duck Cluster
M26	Scutum	18	45.2	−09	24	8.0	15	Binoculars
NGC 2353	Monoceros	07	14.6	−10	18	7.1	20	60
NGC 2506	Monoceros	08	00.2	−10	47	7.6	7	60
M46	Puppis	07	41.8	−14	49	6.1	27	Binoculars
NGC 2360	Canis Major	07	17.8	−15	37	7.2	13	50

Object	Constellation	RA h	RA m	Dec °	Dec ′	Mag	Size arc min	Comments
M23	Sagittarius	17	56.8	−19	01	5.5	27	Binoculars
M25	Sagittarius	18	31.6	−19	15	4.6	32	Naked eye
M41	Canis Major	06	47.0	−20	44	4.5	38	Naked eye
M21	Sagittarius	18	04.6	−22	30	5.9	13	Binoculars
NGC 2362	Canis Major	07	18.8	−24	57	4.1	8	60
NGC 2451	Puppis	07	45.4	−37	58	2.8	50	Naked eye
M6	Scorpius	17	40.1	−32	13	4.2	15	NE; Butterfly Cluster
M7	Scorpius	17	53.9	−34	49	3.3	80	Naked eye
NGC 2477	Puppis	07	52.3	−38	33	5.8	27	Bins; southern showpiece
NGC 6124	Scorpius	16	25.6	−40	40	5.8	29	Binoculars
NGC 6231	Scorpius	16	54.0	−41	48	2.6	15	Binoculars
NGC 5823	Circinus	15	05.7	−55	36	7.9	10	60
NGC 3532	Carina	11	06.4	−58	40	3.0	55	Naked eye
NGC 4755	Crux	12	53.6	−60	20	4.2	10	50; Jewel Box Cluster
NGC 2516	Carina	07	58.3	−60	52	3.8	30	Naked eye
NGC 3766	Centaurus	11	36.1	−61	37	5.3	12	Naked eye
IC 2602	Carina	10	43.0	−64	24	1.9	50	NE; Southern Pleiades

NEBULAE

Object	Constellation	RA h	RA m	Dec °	Dec ′	Mag	Size arc min	Comments
NGC 7023	Cepheus	21	01.8	+68	10		18 × 18	150; surrounds a star
NGC 7635	Cassiopeia	23	20.7	+61	12		15 × 8	100; Bubble Nebula
IC 1396	Cepheus	21	39.1	+57	46		170 × 130	150
IC 5146	Cygnus	21	53.5	+47	16		12 × 12	75; Cocoon Nebula
IC 405	Auriga	05	16.2	+34	16		30 × 19	200; Flaming Star Nebula
NGC 2264	Monoceros	06	37.6	+09	54		10 × 7	150; Cone Nebula
NGC 2261	Monoceros	06	39.2	+08	44	9	2 × 1	60; Hubble's Variable Nebula
M78	Orion	05	46.7	+00	03	8	8 × 6	60
NGC 2024	Orion	05	41.9	−01	51	30	7	5; Flame Nebula
M42	Orion	05	35.4	−05	27	4	66 × 60	Naked eye; Orion Nebula
M16	Serpens	18	18.8	−13	47		21	Naked eye; Eagle Nebula
M17	Sagittarius	18	20.8	−16	11	7	46 × 37	Binoculars; Omega Nebula
M20	Sagittarius	18	02.6	−23	02	8	20 × 20	60; Trifid Nebula
M8	Sagittarius	18	03.8	−24	23	6	90 × 40	Naked eye; Lagoon Nebula
NGC 3372	Carina	10	43.8	−59	52		120 × 120	NE; Eta Carinae Nebula
NGC 2070	Dorado	05	39.0	−69	10		40 × 25	NE; bright spot in LMC

SUPERNOVA REMNANTS

Object	Constellation	RA h	RA m	Dec °	Dec ′	Mag	Size arc min	Comments
NGC 6960/92	Cygnus	20	56	+31	43		70 × 6	100; Veil Nebula
IC 443	Gemini	06	17.9	+22	47		50 × 40	150
M1	Taurus	05	34.4	+22	01	9	6 × 4	60; Crab Nebula

GLOSSARY

achromat A lens that corrects to some extent for the serious *false color* seen in the image produced by a single lens. It usually consists of two lenses of different types of glass combined in a cell.

altazimuth mount A telescope mount with separate axes for moving in *altitude* and *azimuth.*

altitude In astronomy, angular distance above the horizon in degrees.

aperture The clear diameter of a telescope's main mirror or lens

apochromat A lens designed to overcome almost completely the problem of *false color*, with better correction than an *achromat.*

azimuth Angular distance around the horizon, measured from north (0°) through east (90°), south (180°) and west (270°).

Barlow lens A lens that increases the working focal length of an optical system. Placed before the eyepiece, it increases the magnification of that eyepiece.

Cassegrain A reflecting telescope system with a secondary mirror that reflects the converging beam of light back toward the primary mirror. A hole in the primary mirror allows the image to be viewed from the back of the telescope.

catadioptric An optical system that combines both lenses and mirrors to form an image. The lens is designed to correct for imperfections in the image provided by the mirror.

coma An imperfection in an optical system that results in star images some way from the center of the image appearing not as points but with short tails pointing away from the center.

coma corrector An optical device for overcoming the *coma* in an image. It is placed in the light path near the eyepiece.

declination The position coordinate of an object in the sky that corresponds to latitude on Earth. The scale runs from +90° at the north celestial pole to 0° at the equator and −90° at the south celestial pole.

dew cap An extension of a telescope tube that restricts the exposure to the night sky of the telescope's lens or corrector plate, thus helping to slow down the accumulation of dew on its surface.

diffraction An effect caused when light slightly bends round any obstacle in its path, such as the aperture of a telescope or the vanes of the spider that holds the secondary mirror. It gives rise to a false disk in the image of a star, and spikes on the star image, and also limits the *resolving power* of the telescope. Diffraction-limited optics should have their performance limited by diffraction rather than by errors in manufacture.

Dobsonian An *altazimuth* telescope mount designed for simplicity of construction, in which the telescope tube pivots in altitude on a box which rotates in azimuth about a central pivot. The classic design uses Teflon pads to provide low friction bearings.

driven mount A telescope mount with motor drives to move it in *right ascension* and, often, *declination.* The RA motor is set to drive the telescope at *sidereal rate* in order to follow the stars without further correction, while the declination motor is designed to allow for small corrections using a push-button handset.

ED (Extra-low Dispersion) A special type of glass used to make lenses that are highly corrected for *false color*.

Erfle A type of eyepiece that gives a wide apparent field of view. Usually available in 20 mm focal length.

equatorial mount A telescope mount with one axis parallel to the Earth's axis, thus enabling the Earth's daily rotation to be counteracted by a single movement.

false color Colored fringing, correctly known as chromatic aberration, around the edges of objects in an image produced by a lens. It is caused by the glass of the lens splitting light into its component colors, which come to different focus points.

finder A small telescope with a wide field of view and crosswires, mounted on a telescope to assist in pointing the main instrument at a chosen object.

fluorite lens A lens that includes one component made from fluorite crystal, which makes possible very good correction for *false color*.

focal length The distance between a lens (or mirror) and the image it produces of an object at infinity.

focal ratio The ratio between a lens or mirror's focal length and its diameter, usually written as, for example, f/8 for a lens whose focal length is eight times its diameter.

focal reducer An optical device placed in the light beam of a telescope before it reaches the eyepiece which reduces the focal ratio of the system. It has the opposite effect from a Barlow lens.

German equatorial mount (GEM) A telescope mount with two shafts, one of which (the polar axis) is set parallel to the Earth's axis and carries the other (the declination axis). This in turn carries the telescope at one end with a counterweight at its other end.

Go To A means of directing a telescope to a chosen object in the sky using computer control.

King rate A rate for driving a telescope that takes into account the average refraction of the atmosphere, thus giving slightly improved star tracking.

LPR filter (light-pollution rejection filter) A filter designed to cut down light pollution.

Maksutov A *catadioptric* reflecting telescope system in which the imperfections of the image from a spherical mirror are corrected by a plate at the top end of the telescope tube. The corrector plate of a Maksutov is a steeply curved piece of glass. It is also known as a Maksutov-Cassegrain, as the light is usually reflected from the secondary through a hole in the main mirror. Maksutov-Newtonians are also possible, in which the light is reflected by a flat secondary to the side of the tube to give a shorter focal ratio than with the Maksutov-Cassegrain.

Nagler An eyepiece design that gives a very wide apparent field of view, designed by Al Nagler and marketed by Tele Vue Optics.

Newtonian The simplest reflecting telescope design, with a concave mirror to collect light and a secondary mirror at the top end of the tube just before the light comes to a focus, to reflect the image to the side of the tube where it can be seen.

objective (OG) The main lens of an optical system. The term does not usually refer to the main mirror. It is sometimes abbreviated OG, for object glass.

orthoscopic An eyepiece design that gives a flat field of view with no distortions.

OTA (optical tube assembly) The tube, main optics and focusing mount of a telescope, without eyepieces or accessories.

parabolic The ideal shape in cross-section of a telescope mirror. Small mirrors of long focal ratio may have a spherical cross-section without significant loss of performance.

periodic error correction (PEC) An electronic system for correcting known mechanical errors in the worm and wheel used to drive telescopes.

Plössl A design of eyepiece particularly suited to telescopes with low focal ratio, and now virtually the standard for all instruments.

push to A manual equivalent of the computerized Go To telescope pointing system.

relay lens A lens used to transfer an image from one position to another.

resolution, resolving power The ability of an optical system to discriminate fine detail in an image. Larger mirrors or lenses have a better resolving power than smaller ones.

right ascension, RA The celestial equivalent of longitude on the Earth. It is usually measured in hours and minutes because it is related to the daily turning of the Earth.

Schmidt-Cassegrain (SCT) *Catadioptric* telescope system with a spherical primary mirror of very short focal length, a corrector plate at the top of the tube to overcome the optical defects of a short-focal-length spherical mirror, and a secondary mirror that sends the light down through a hole in the main mirror to the eyepiece.

seeing The effect of turbulence in our atmosphere on the image seen through a telescope, causing it to be blurred and in motion.

setting circles Engraved or printed circular scales on the axes of an equatorial mount that allow the telescope to be pointed at any chosen point whose coordinates are known. Separate circles are needed for the RA and declination axis.

sidereal rate The rate at which stars move through the sky as a result of the Earth's daily rotation.

spider Astronomically, the vanes that hold the secondary mirror in a reflecting telescope.

telecompressor Another name for *focal reducer.*

UHC filter (Ultra High Contrast filter) A light-pollution filter which transmits a band of spectrum that includes the emission from most nebulae while cutting out the rest of the spectrum, over which light pollution dominates.

INDEX

Page numbers in *italics* refer to illustrations or their captions

Acknowledgements

Thanks to the following manufacturers for their help in supplying photographs, information and test instruments for this book: Telescope House for Meade (www.telescope-house.com); David Hinds Ltd for Celestron (www.dhinds.co.uk); Green Witch (www.green-witch.com) for telescope mounts (page 55); Losmandy Optical Products (www.losmandy.com); Optical Vision Ltd for Sky-Watcher/Synta and TAL (www.opticalvision.co.uk); Orion Optics for Orion Optics and Vixen (www.orionoptics.co.uk); Orion Telescopes and Binoculars (www.telescope.com); Parks Optical (www.parksoptical.com); Tele Vue Optics, Inc. (www.televue.com).

Additional photographs by page number except where otherwise credited: **7** Howard Brown-Greaves; **106** (top) David Cortner; **112** (inset), **124–5** Damian Peach; **137** (top), **138** (bottom), **139** Michael Stecker; **140** Nik Szymanek; **141** (top) Michael Stecker; **163** Dave Tyler. All other photographs and illustrations by the author.